Android, Assembled

Steve Jones
General Editor

Vol. 126

Android, Assembled

The Explicit and Implicit Anatomy of Social Robots

Edited by
Jaime Banks

PETER LANG
New York - Berlin - Bruxelles - Chennai - Lausanne - Oxford

Library of Congress Cataloging-in-Publication Control Number: 2024031333

Bibliographic information published by the Deutsche Nationalbibliothek.
The German National Library lists this publication in the German
National Bibliography; detailed bibliographic data is available
on the Internet at http://dnb.d-nb.de.

Cover design by Peter Lang Group AG

Cover Art by Katerina Belikova, Artist, 2024.
Printed by permission of the artist. https://www.instagram.com/ninjajo_art/

ISSN 1526-3169 (print)
ISBN 9781636672038 (paperback)
ISBN 9783034352970 (hardback)
ISBN 9781636672014 (ebook)
ISBN 9781636672021 (epub)
DOI 10.3726/b22246

© 2025 Peter Lang Group AG, Lausanne
Published by Peter Lang Publishing Inc., New York, USA
info@peterlang.com - www.peterlang.com

All rights reserved.
All parts of this publication are protected by copyright.
Any utilization outside the strict limits of the copyright law, without the permission of the publisher, is forbidden and liable to prosecution.

This applies in particular to reproductions, translations, microfilming, and storage and processing in electronic retrieval systems.
This publication has been peer reviewed.

Contents

List of Figures   ix
List of Tables   xi
Introduction: (Dis)Assembling the Android   1
    Jaime Banks

Part I: Explicit Anatomy

Chapter One: Morphology & Ontology: The Convergence of Form and Being   11
    Autumn Edwards

Chapter Two: Heads & Faces: An Assemblage of Cues   23
    Xiaoyu Jia & Chien-Hsiung Chen

Chapter Three: Eyes & Gaze: Exchanging Glances with Robots   33
    Chris Chesher

Chapter Four: Legs & Feet: Complexities of Limb-Environment Interactions   43
    Zhenyu Gan

Chapter Five: Wings & Propellers: Caring with/for Flying Robots   53
    Mafalda Gamboa

Chapter Six: Color & Clothing: The Social Consequences of Aesthetics   63
    Natalie Friedman

Chapter Seven: Tactility & Texture: Embrace at the Interface   73
    Jason Edward Archer

Chapter Eight: Gesture & Posture: Conveying Simulated Emotional States   85
    Sara Ali

Chapter Nine: Text & Speech: Robot Communication Design   95
    Kristine L. Nowak & Antonio Chella

Chapter Ten: Screens & Links: Playful Affordances of Future Friends   105
    Katriina Heljakka

Chapter Eleven: Memory & Information: The Core of Robot Cognition   115
    Rafael Sousa Silva & Tom Williams

Chapter Twelve: Sensors & Actuators: Synergies of Perception and Movement   125
    Uchenna Ogenyi

Chapter Thirteen: Implants & Injections: Long-Lived Integrations with the Organic   135
    Dayeoun Jang & Stephanie Jordan

Chapter Fourteen: Aggregation & Distribution: Beyond the Singular Form   145
    Sarah Diefenbach, Daniel Ullrich, & Andreas Butz

Part II: Implicit Anatomy

Chapter Fifteen: Images & Frames: Tensions in Representations   157
    Aike C. Horstmann

Chapter Sixteen: Digitality & Interactivity: Lessons Learned from NPCs   167
    Nicholas David Bowman, Elena Yifei Zhao, & Yoon Esther Lee

Chapter Seventeen: Cuteness & Repulsiveness: Aesthetics of Machine Bodies   179
    Joel Gn

Chapter Eighteen: Sex & Gender: A Complicated Relationship   187
    Leopoldina Fortunati

Chapter Nineteen: Power & Agency: A Dynamic Interplay   197
    J. Nan Wilkenfeld

Chapter Twenty: Authority & Status: Mechanisms of Influence   207
    Tomasz Grzyb & Dariusz Doliński

Chapter Twenty One: Membership & Roles: Complicating the Notion of "Teaming" with Machines   217
    Qingyu Liang

Chapter Twenty Two: Cognition & Context: Closing the Social Gap   227
    Roc Myers
Chapter Twenty Three: Decision & Action: A New Kind of Assemblage   237
    Sarah Rajtmajer
Chapter Twenty Four: Life & Death: Making Sense of Robots' Temporary
    Presence   247
    Kevin Koban
Chapter Twenty Five: Mind & Morality: Seeing the Ghost in the Shell   255
    Jan-Philipp Stein
Chapter Twenty Six: Sword & Shield: Do Robots Have Defensive Obligations? 265
    Nicholas G. Evans
Chapter Twenty Seven: Persons & Things: Rethinking the Ontology of the
    Robot   273
    David J. Gunkel

Index of Social Robots   283
Notes on Contributors   291
Index   301

List of Figures

Figure 03.1: A 1914 British military recruitment advertisement depicting British Secretary of State for War; the robot "Pepper" by SoftBank Robotics. 36
Figure 05.1: A set of images with a focus on drones and their propellers. 58
Figure 07.1: Living AI's EMO in bear robe designed by Ara Archer Mira. 76
Figure 10.1: "Ron" the B*bot, a speculative IoToy. 111
Figure 14.1: Prototypical sketching and experimentation with possible elements of the room intelligence (RI). 150
Figure 14.2: Characterizing familiar complex systems by their degree of physical distribution and their degree of control distribution. 153

List of Tables

Table 21.1: Insights from human teaming literature for understanding human-AI teaming constructs. 220

Table 21.2: Factors that may influence people's perception of AI as a teammate. 223

Introduction: (Dis)Assembling the Android

JAIME BANKS

When I was small, my parents brought into our home a Teddy Ruxpin—an animatronic, humanoid bear-like being that could speak and move. From the box and batteries installed, he blinked his eyes open and in a rather soft and kindly voice asked (as he did for millions of children): "Can you and I be friends?" As he spoke his plush snout moved and eyes blinked. He wore a tan-colored smock that hid some of the bits that would make him tell stories and sing: A cassette-tape deck built into his back and a speaker constituting the bulk of his belly. His behaviors would match the content of dozens of exchangeable cassette tapes, including the sounds of his own voice, sounds of other characters that we were to imagine in those stories, and both diegetic sounds and dramatic music to set the scene, all adjusted with a volume wheel. Apparently, unknown to five-year-old me, those tapes also carried a set of inaudible cues embedded in certain tracks that would trigger the movements (see With or Without Fur, 2018). Those movements were based on three servo motors (one each for the eyes and mouth) and made appealing by a furry body that encased some other components to give him shape and transform the tape signals to observable behavior (see What's Inside?, 2016). Batteries, eyes, face, voice, clothes, speaker, belly, tapes, sounds, controls, motors, fur, shape, signals, and movements were assembled to manifest what was, to me, a bit of magic… at least until my brother took it swimming in the kiddie pool.

ROBOTS AS SOCIOTECHNICAL ASSEMBLAGES

In starting to write this introduction, I was surprised at how many of that robot's particulars I could recall—the slowness of its blink, the tediousness of exchanging the cassettes, the bulkiness of its body—only because we often think about and talk about "robots" as singular and whole thing rather than as a collection of parts. The objective of this book is to explore that tension by considering the components of social robots both in terms of what social robots manifestly *are* and what we humans *read into* them—their explicit structural and operational elements and the elements we interpret in their structures and operations.

To this end, the notion of an assemblage is a useful one. In this characterization, I draw primarily from Nail's (2017) formalization of Deleuze and Guattari's (1983, 2008) assemblage theory, though there is some kinship with DeLanda (2006) and Latour's (2005) works. *Assemblage* refers to a construction or arrangement of assorted things that are held together by networked external forces, and that arrangement exists as a multiplicity rather than as a unity; a multiplicity is contingent on its composition and composing conditions at a particular point in time. The name we give to an assemblage is more a name given to the force-machinery—the set of conditions—by which its elements are arranged. Robots, broadly, as a category of thing is a reference to the types of arrangements typical of things we recognize as robots. The proper noun "Ray" (the Robothespian in my research laboratory) refers to a particular arrangement of things that follows the prototypical arrangement to an extent but is different from other arrangements—Ray is itself and no other robot, remains Ray so long as the forces and constituted arrangement are similarly sustained so as to be recognizable, and will cease to be Ray when the arrangement is otherwise. That other-arrangement could come as the former-Ray might be altered by me, unplugged by a maintenance worker, broken by circumstance, or otherwise rendered different.

Assemblage is a useful tool to think about what robots are and what they mean to the world as actually or interpretably social in (at least) three ways. First, the framework (as with kin approaches) requires that we decenter human agency so we can more fully consider the way that more-than-human elements and immaterialities function as arrangement elements, alongside the human and the concrete; this can animate richer discussions of how the robot-assemblage manifests sociality independent of us and among its constitutive components, or perhaps in forms of sociality that we may not even recognize (see Bogost, 2012). Notably, this does not require discounting human *experience* as we may be among the arranging forces constituting robots, may be co-assembled with robots, or may even be assemblages entangled with robotic elements as with cyborgs or as part of higher-order assemblages.

Second, it creates space to acknowledge the independence, agency, and meaning of parts. Robots fit Nail's machinic metaphor for a multiplicity: Unlike a human body in which the organs work together and reproduce their co-relations, the components of a robot may be discrete. As he notes, a heart taken from a human body does not survive as a heart and the body does not survive without it; however, it may be that a robot relieved of its head could potentially still function as both head and a trunk. Notably, this is a challenging mode of thinking, as we are accustomed to our embodied way of being in the world and to our egos as the center of experience. But robot parts and their arrangements are key in the way that robots are represented in media, designed in labs, engaged as interlocutors, and regulated by norms and law. Take, for instance, any prototypical humanoid and consider the potential to give it genitalia. Separately, the robot (and its arranged components) and the genitalia, separately, might be understood as a machine and a sex toy. But arranged in a particular configuration and embedded in particular cultural contexts, they together become a "sex robot" that conjures moral and normative emotions, may arouse or repulse, and may instigate calls for regulation; when further arranged with other capacity-affording components like decision-making AI that imply a potential for a robot to consent (or not) the contextualized robot-assemblage may then motivate calls for its protection and rights (see Gersen, 2019; Szczuka et al., 2019).

Finally, the notion of assemblage can animate attention to the forces that maintain the arrangement—that is, what pressures create and sustain and degrade the arrangement of a robot's parts and, in the entanglement of human and machine, what pressures do the same for a robot's ostensible sociality? There are myriad forces that may be at play. Contenders include material and functional forces: How do properties of a robot's constitutive physical parts and their co-related workings contribute to how they do (not) remain arranged? There are normative and cultural forces: What is considered proper or customary or natural or legal in the arrangement—and what is not? Scientific and technological forces may be at play: What arrangements are (not) possible given the state of human knowledge and of mastery over materials and processes? Considering social systems, to what extent and nature are robotic arrangements subject to economic and political forces? How do, for instance, capitalist structures support particular (dis)arrangements and how might elements be arranged differently in other structures? There are psychological and physiological forces as well: How is our imagination for arrangements linked to our own experiences of the world? And how do our embodied reactions to some arrangements make them more or less desirable to replicate? And what of temporal and spatial forces: As multiplicities, to what extent are robotic arrangements "nomadic" versus more stable—changing, reconfiguring, or dissolving component-relations across time and context? Some of these questions are addressed in the forthcoming chapters, and some can be more tacitly found in the cracks between them.

Of note, the application of assemblage frameworks to robots is not a new one. They've been engaged as frames for understanding how humans and robots may exist in reciprocal relations (a potential often ignored as we focalize human ways of benefiting; Kerruish, 2021), to consider how human imagination functions as part of robotic assemblages (as opposed to being a function of imagination; Sumartojo et al., 2021), and to argue for potential multi-species assemblages that may require us to consider what *kind* of thing a robot is according to how it relates to other things (versus what it manifestly is; Kubes & Reinhardt, 2022). Some robots may be configured as communicative arrangements of machine geographies and individual experiences (themselves complex assemblages) held together by flows of human affect and attraction (Dehnert, 2022). There are self-assembling and self-repairing robots that, according to their design, have internal forces that maintain or reconfigure their own arrangements (e.g., Saintyves et al., 2024) and multi-robot systems that arrange themselves into higher-order assemblages (see Bray & Groß, 2023). Robotic assemblages contribute to higher-order assemblages as a function of their arranged parts, as when robotic guide dogs may co-construct "vision" with its human partner (Due, 2023).

Beyond these and similar works, the present volume illuminates the discrete parts that tend to be assembled in archetypical social robots and challenges us to examine those elements and the forces that bind them—how the parts and the whole contribute to the social that we tend to associate with explicitly or implicitly agentic robots. The charge to you, readers, is to more broadly consider: In what ways is a social robot limited to or more than the sum of these parts and forces?

THIS VOLUME

The framing and questions above are not all that distinct from those posed in this book's predecessor that focused on videogame avatars (see Banks, 2018), but differ in a key way: The focal machines have a physical body (rather than digital) and are often autonomous to some extent (rather than controlled, by definition). This means the answers to those questions often take a different track. Social robots can start to look and feel more like other machines embedded in human social life and even more like humans themselves, so some of the considerations of components and forces from other domains of life can be replicated in robotic arrangements. This book's title was originally to be *Agent, assembled* to leave room for a range of robot morphological to be addressed. But interestingly nearly all chapters gravitated toward human-like characteristics or components, theories of anthropomorphism or humanlikeness interpretations, or applying human social concepts to nonhuman components (e.g., Chapter Five that discusses how we may engage notions of care in relation to robots with wings and propellers). Notably, some chapters also challenge anthropocentric defaults, considering the ways robotics can be otherwise shaped, distributed, adorned, or animated.

In engaging authors to contribute to this book, I offered them a challenge: For each pair of components, summarize the state of science or philosophy and then ... *provoke*. You'll find the former explains where we've been and where we are now while the latter often speculates as to where we're going. Some of those provocations are lighter and perhaps even fun to think about—for instance, does a robot need clothes as a human does or might its apparel needs be different? Others are heavier and may resonate with hopes or fears you have around robots—should robots be made to participate in warfare? Across that spectrum, I hope you'll find claims and questions to animate your thinking about what robots are and what they may become, and how their discrete components can play an important role in those actuals and potentials.

The chapters in this volume are organized into two parts: In the first are more explicit bits of anatomy—the parts and pieces that physically and digitally manifest the robot-as-thing. In Chapter One, Edwards sets us on a path to discover the whole and parts of robots through the lens of robot morphologies of robots—and how, despite the assembled nature, we often perceive them as a unified whole. In Chapters Two and Three, Jia, Chen, and Chesher emphasize the head—an assemblage in itself—as a centerpiece of our perception of robots as mindfully social agents, as they look to us and elsewhere. In Chapters Four and Five we consider robot appendages—from the more anthropo- and zoomorphic legs and feet as they interface with surfaces to the less familiar wings and propellers as they interface with the air and sometimes with us—courtesy Gan and Gamboa. Then we move into questions of surfaces and how they are adorned, where Friedman and Archer in Chapters Six and Seven prompt us to consider whether touch and aesthetics need necessarily mirror the social and material dynamics of organic bodies. Chapters Eight, Nine, and Ten by Ali, Nowak, Chella, and Heljakka emphasize the technical and social dynamics of social robots' encoding of meaning, animating questions about the authenticity or artificiality of our connections with them. Silva, Williams, and Ogenyi—in Chapters Eleven and Twelve—unpack the way social robots apprehend, store, and process information, and whether our current methods will necessarily endure as best-fit as social robotics advances. Then in Chapters Thirteen and Fourteen, Jang, Jordan, Diefenbach, Ullrich, and Butz challenge us to go beyond the notion that a robot must be a singular and unified body, but perhaps incorporated into our own bodies or distributed across space.

In the second half are more implicit elements—those that we tend to interpret or *read into* the machine as a function of the explicit. Chapters Fifteen and Sixteen attend to influences and insights from popular media, as Horstmann details the influences of robot representations on how we response to actual machines and Bowman, Zhao, and Lee invite us to consider the insights from gaming psychology as it may inform human-robot interaction studies. We are charged with considering how we apply human norms and categories to social robots in Chapters Seventeen and Eighteen, as Gn suggests cuteness is a way to negotiate gaps between us and robots while Fortunati explores the interplays between human and robot genders. Chapters Nineteen, Twenty, and Twenty One see Wilkenfeld, Grzyb, Dolinski, and Liang unpacking the ways that social robots are influential as they interact with us individually and in groups. Two different takes on robot cognition are offered by Myers and Rajtmajer in Chapters Twenty Two and Twenty Three, where the former points to the role of context in constituting shared meanings and the latter examines how generative AI may shift our models for thinking about mindful action. From Koban and Stein come Chapters Twenty Four and Twenty Five; they consider how we see a mindful *someone* in the machine, and whether or how that someone can be said to live or die. Finally, Chapters Twenty Six and Twenty Seven round out the exploration of implicit anatomy by asking whether and how social robots are a kind of thing that has certain kinds of responsibilities.

At the end of the book, you will find an index of robots, with additional information on robots mentioned in the chapters. Notably, with the pace of development in robotics and AI, there is a possibility that this information will be outdated by the time you read this. New robots will emerge as better exemplars, discussed robots will be sunset by their companies or become otherwise obsolete, and the cutting-edge functionality of today's machines will be eclipsed. That index, then, offers some context for the robots discussed here but also serves as a snapshot of relevant social robots at the time of publishing—an important time (largely driven by the recent mainstreaming of generative AI) when industry, academia, and the public more broadly are discussing with seriousness the extent to which robots can be like humans.

REFERENCES

Banks, J. (Ed.). (2018). *Avatar, assembled: The social and technical anatomy of digital bodies*. Peter Lang.
Bogost, I. (2012). *Alien phenomenology, or, what it's like to be a thing*. University of Minnesota Press.
Bray, E., & Groß, R. (2023). Recent developments in self-assembling multi-robot systems. *Current Robotics Reports, 4*, 101–116.

Dehnert, M. (2022). Sex with robots and human-machine sexualities: Encounters between human-machine communication and sexuality studies. *Human-Machine Communication, 4*, 131–150.
DeLanda, M. (2006). *A new philosophy of society: Assemblage theory and social complexity.* Continuum.
Deleuze, G., & Guattari, F. (2008). *A thousand plateaus: Capitalism and schizophrenia* (B. Massumi, Trans.). Continuum.
Deleuze, G., & Guattari, F. (1983). *Anti-Oedipus: Capitalism and schizophrenia.* Trans: R. Hurly. University of Minnesota Press.
Due, B. L. (2023). Guide dog versus robot dog: Assembling visually impaired people with non-human agents and achieving assisted mobility through distributed co-constructed perception. *Mobilities, 18*(1), 148–166.
Gersen, J. S. (2019). Sex lex machina. *Columbia Law Review, 119*(7), 1793–1810.
Kerruish, E. (2021). Assembling human empathy towards care robots: The human labor of robot sociality. *Emotion, Space and Society, 41*, 100840.
Kubes, T., & Reinhardt, T. (2022). Techno-species in the becoming towards a relational ontology of multi-species assemblages (ROMA). *Nanoethics, 16*, 95–105.
Latour, B. (2005). *Reassembling the social: An introduction to Actor-Network Theory.* Oxford University Press.
Nail, T. (2017). What is an assemblage? *SubStance, 46*(1), 21–37.
Saintyves, B., Spenko, M., & Jaeger, H. M. (2024). A self-organizing robotic aggregate using solid and liquid-like collective states. *Science Robotics, 9*(86).
Szczuka, J. M., Krämer, N. C., & Hartmann, T. (2019). Negative and positive influences on the sensations evoked by artificial sex partners: A review of relevant theories, recent findings, and introduction of the Sexual Interaction Illusion Model. In Y. Zhou & M. H. Fischer (Eds.), *AI love you: Developments in human-robot intimate relationships* (pp. 3–19). Springer.
Sumartojo, S., Lundberg, R., Tian, L., Carreno-Medrano, P., Kulić, D., & Mintrom, M. (2021). *Geoforum, 124*, 99–109.
What's Inside? (2016). What's inside THE ORIGINAL TEDDY RUXPIN? YouTube. <https://www.youtube.com/watch?v=sJ5cD9xvqew>.
With or Without Fur. (2018). How to make a 1985 Teddy Ruxpin say anything you want. YouTube. <https://www.youtube.com/watch?v=eGVB5SdGstM>

PART I

Explicit Anatomy

CHAPTER ONE

Morphology & Ontology: The Convergence of Form and Being

AUTUMN EDWARDS

> DOMIN.: ...You see with the help of his tinctures he could make whatever he wanted. He could have produced a Medusa with the brain of Socrates or a worm fifty yards long— (*She laughs. He does also; leans closer on couch, then straightens up again*) —but being without a grain of humor, he took into his head to make a vertebrate or perhaps a man. (Čapek, 1920/2004, 1.1.14–15)

In 1920, the term "robot" made its debut in Karel Čapek's stage play *R.U.R.* (Rossum's Universal Robots). In this first instance, the term referred to organic humanoids created to serve as laborers for their human creators. The opening scene of the play featured a dialogue considering the profound questions: What is a man? What is a robot? and What is their difference? These inquiries laid the foundation for ontological exploration—the philosophical inquiry of existence and being. The answers to these ontological questions[1] inevitably shaped robots' morphologies—that is, their physical forms. This intricate relationship between ontology and morphology in robot design is the central theme we will explore.

---

1 In the fictionalized dialogue in *R.U.R.*, the original answer was that a robot must be shaped in accordance with its category of being and how its category relates to others, for ontology is never understood in isolation but only by a process of referring and deferring to other types of being with which we are familiar.

When we encounter a robot, our initial question is often: What is this? The robot's outward appearance provides us with visual cues, allowing us to distinguish it from its physical surroundings and decipher its type of being and role. This visual information helps us establish a connection between the robot and our existing knowledge and experiences, enabling us to assess its similarities and differences in relation to entities we are already familiar with (Edwards, 2018). The process of understanding a robot typically progresses from prototyping (attempting to grasp its essence) to stereotyping (formulating assumptions about its behavior, drawing on behaviors of known exemplars) to the construction of a personalized profile that shapes our impressions and interaction expectations, especially with robots designed for social engagement (Edwards et al., 2019). This visual analysis is not merely an incidental step but a critical one, as human cognition is deeply iconophilic. We often conjure up mental imagery, even of unseen robots, based on descriptive language, which underscores the integral role of a robot's appearance in our conceptualization of its identity and its potential impact on our lives. Considering this, we can assert that the initial step in comprehending any entity, particularly one that is animate or social, is to provisionally prototype or ontologize it on the basis of its outer form.

The interplay between form and being is not unique to the realm of robotics. In philosophy, ontology delves into the essence of being, guiding us in the classification and comprehension of various entities, whether human, natural, or synthetic (Taylor, 1959). Similarly, in art and literature, the form is more than mere structure; it shapes our understanding of a work's very essence (Thomasson, 2005). The same is true in architecture and design, where the tangible shape of structures and objects conveys deep cultural, social, and functional narratives integral to their identities (e.g., Blier, 1994). Acknowledging this wider scope, we can begin to grasp that the entanglement of form and being poses significant questions across different spheres, challenging us to rethink our perception of, categorization of, and engagement with entities.

In this context, it is essential to consider the human cognitive processes that underlie our understanding of robots' forms and ontologies. I argue that our perception of robots involves a gestalt-analytical approach, akin to how we perceive other entities upon first encountering them (Palmer, 1999). Initially, we view the robot as a whole—a gestalt—to swiftly categorize them through recognizable patterns and associations, forming a preliminary ontological framework. This initial, holistic perception sets the stage for deeper understanding, where we shift to analytical processing: Deconstructing the robot into its components and evaluating the significance of each, which contributes again to the larger, unfolding gestalt. This dynamic interplay between gestalt and analysis reflects our innate cognitive strategies and highlights the importance of design in robot morphology and ontology.

This recognition that the cognitive processing of a robot is not a linear pathway but rather an oscillation between the holistic and the particular also allows for a more nuanced appreciation of robotic form. From this perspective, we understand that the silhouette of a robot, its textural surface, the articulation of its limbs, and even the color scheme, when combined, form a gestalt that is greater than the sum of its parts. (See Chapters Six to Eight for more on those elements, separately.) These elements collectively influence our ontological sense-making as we endeavor to understand what a robot is and how it fits into our conceptual framework of beings. Each design element thus becomes a lens through which we interpret the robot's intended role and capabilities.

UNDERSTANDING ROBOT MORPHOLOGY: SHAPING PERCEPTIONS

In the world of robotics, the term "morphology" encompasses myriad design elements that collectively define a robot's physical form. These components are not merely aesthetic considerations but rather the building blocks of a robot's identity in the world and the early prototyping information from which we surmise how it interacts with its environment and how to perceive it. Robot morphology comprises various modular components and how they are linked to the ontologizing process in which people categorize and define the essence of robotic beings.

Silhouette and Body Structure

In robotic design, the silhouette and overall body structure are fundamental cues for deciphering a robot's nature, capabilities, and intended use. Broadly speaking, designs range from anthropomorphic (human-like) and zoomorphic (animal-inspired) to phytomorphic (plant-like), theomorphic (deity-like), mechanomorphic (machine-like), and creaturely (fantastical) forms. We might interpret a humanoid figure as a sign the robot was designed for social interaction, while we interpret a mechanical form as a practical tool (Kwon et al., 2016). These holistic silhouettes serve as an initial framework within which more granular variations exist. For instance, within the anthropomorphic category, robots can be modeled to echo the forms of different human identities—men, women, or children—or to reflect diverse groups characterized by age, ethnicity, and other distinct features. These evocative details may tap into our mental models of familiar beings with the potential to deeply affect our interactions with and reactions to these robots (Bernotat et al., 2017; Strait et al., 2018). This is one reason that robot body structure is often biomimetic, emulating natural designs and processes to elicit recognition. The size of the robot also interplays with its silhouette to inform our understanding. For example, larger robots (like the *mecha*, or giant robot, Kuratas) might be perceived as more capable, powerful, or anxiety-provoking, while smaller ones (like Anki's Cozmo or SoftBank Robotics' Nao) could be seen as more accessible or benign (Hiroi & Ito, 2008). Through these layered design choices and the recursive interplay between the general and specific, our conceptualizations of robots take shape.

Mobility and Locomotion

Mobility and locomotion choices are also significant components of robot morphology that contribute to our sense-making. A robot's capacity to move independently or not and its methods of travel and navigation (e.g., walking, rolling, flying, or swimming) reveal how it interacts with its environment, affecting our perception of its role and category within the robotic spectrum. A robot equipped with wheels might be categorized as a vehicle designed for efficient transportation, while one with legs could be seen as an entity capable of agile, human-like movement. The number of locomoting features may also influence perceptions of a robot; for example, a spider-legged robot in movement may be perceived as more aggressive compared to robots moving in other ways (Sims et al., 2005). Such design choices directly impact how we conceptualize the robot's purpose and place in our world and its social potential.

Manipulators and End Effectors

The design of a robot's manipulators and end effectors is crucial for its functional role in various contexts. Manipulators are analogous to the limbs of a robot, appendages composed of segments and joints allowing for movements such as reaching and grasping (Cui & Trinkle, 2021). End effectors, comparable to hands at the end of human arms, are the tools or devices that interact directly with the environment. These can range from simple clamps to intricate instruments designed for specific tasks (e.g., a pollinator robot's brush, vibrating rod, or air blaster used to distribute pollen; Broussard et al., 2023). When robots are equipped with articulated arms and precise end effectors, we tend to view them as capable of performing complex work, perhaps comparable to human assistants or workers. Conversely, robots with more unconventional appendages, or feature atypicality, may be perceived as more eerie or uncanny compared to prototypical designs (Strait et al., 2017). The placement of manipulators and end effectors on a robot also carries the potential to shape perceptions. For instance, Boston

Dynamics' creation of the "Spot Arm," which attaches on top of its head, has sparked both fascination and suspicion among the public (Moses & Ford, 2021).

Material and Texture

The choice of materials and textures used in a robot's construction further guides our ontological interpretation by evoking connections with familiar objects and entities in the physical world. A robot's materials may range from metallic to organic, each choice triggering unique associations and assumptions (see Chapter Six). For example, a robot with soft, human-like skin might elicit associations with organic beings, while a metal-clad robot could be seen primarily as a mechanical creation. The materials and textures chosen for a robot's outer form influence not only our visual interpretation but also our tactile expectations. For example, a robot sheathed in synthetic fur may invite touch in a manner similar to a pet, encouraging a form of interaction that is nurturing and familiar. This tactile quality can create a sense of affinity and warmth, often associated with companion animals. In contrast, a smooth, rigid surface may convey a sense of separation and efficiency.

Adaptability and Modularity

A robot's adaptability and modularity are pivotal in shaping our understanding of a robot's capacity for flexibility and versatility. Adaptability refers to the ability of a robot to adjust its behavior, configuration, or functionality in response to changing circumstances or requirements, enabling it to operate effectively across diverse contexts or environments (e.g., Andriella et al., 2020). Modularity, on the other hand, involves the use of interchangeable components that can be swapped or reconfigured, allowing the robot to perform a diverse range of tasks or to be upgraded with new functionalities over time (Will et al., 1999). From these design choices, we infer a robot's potential to integrate into and evolve within a broader ecology of beings. For instance, modular robots may be a "bucket of stuff" when at rest, but then configure those modules into different forms—cubes, lattices, chains, or biomimetic forms—based on the task at hand (Mackenzie, 2003). In this way, adaptability and modularity inform our ontological assessments, inviting us to perceive the robot as a dynamic and adaptable entity capable of fulfilling multiple roles and functions.

Aesthetics and Theming

Finally, the aesthetic design and thematic elements of a robot significantly inform our perception and classification of it within our social frameworks. The aesthetics of a robot, encompassing its style, shape, and color, often invoke our cultural lexicon, aligning the robot with certain archetypes or narratives familiar to us. For instance, a robot with a sleek, metallic design might be seen through the lens of futuristic innovation, while one with a soft, rounded form may be associated with a more friendly and approachable character. Moreover, color choices can subtly influence our emotional responses and biases; dark hues and red accents might unconsciously evoke the menacing automatons of science fiction lore, triggering a sense of wariness. Conversely, bright, cheery, and warm colors might engender trust and positivity. These design aspects are far from superficial; they are integral to the cognitive process of situating robots within our larger cultural ecosystems, affecting how we relate to and interact with them.

Each of these components of robot morphology influence how we understand their essential nature, and consequently how we interact with and integrate them into our lives. The silhouette, movement capabilities, manipulative functions, material composition, adaptability, and thematic design all coalesce to define a robot's physical form and functions. For this reason, the design of robots requires not just technological sophistication but a careful consideration of the interconnection between form and being.

ROBOT ONTOLOGY: WHERE FORM AND FUNCTION CONVERGE

What is clear so far is that our understanding of a robot is shaped significantly by its physical forms and attributes, especially in our early encounters. To proceed further into the evolving landscape of robot ontology, let's consider two key concepts that illuminate how form and function interact in the domain of perception: The form-function attribution bias and the new ontological category hypothesis.

The Form-Function Attribution Bias (FFAB)

The FFAB highlights the complex relationship between a robot's embodied form and the expectations placed upon it. The FFAB is a systematic cognitive tendency people display when they are perceiving robots. Essentially, this bias arises from an inclination to base perceptions of a robot's functionality on its visual appearance, often attributing functions that may not align with its actual capabilities (Haring et al., 2018). Instead of objectively assessing a robot's functions based on a complete understanding of capabilities, we may judge the robot "by its cover" and anticipate functions rooted in preconceived notions arising from its morphology. For instance, when we encounter a humanoid robot, we may instinctively attribute human-like abilities and consciousness to it, regardless of its actual functionality or lack of subjective experience. This cognitive shortcut is impactful because often our heuristic attributions do *not* match up with robots' actual abilities, which can result in disappointment when the robot falls short of social and functional expectations (Malle et al., 2020). These mismatches between perceived and actual functions ultimately affect various aspects of human-robot interactions, such as trustworthiness, acceptance, and cooperation. As robots become increasingly integrated into society, their acceptance depends, in part, on factors such as the alignment of form and function. The FFAB is a valuable framework for understanding these dynamics (Haring et al., 2018), as researchers gain insights into how morphology influences ontology in human-robot interactions. Understanding this link allows for more informed design decisions, helping to create robots that not only function effectively but also align with users' expectations and perceptions.

A New Ontological Category (NOC)

In the field of Human-Robot Interaction, the NOC suggests that robots occupy a unique ontological category, different from traditional entities. When we encounter robots, we grapple with an unfamiliar form of existence (Weisman, 2022), one that shares kinship with our older, established categories of being in that they are usually clearly machines but also exhibit some of the hallmarks of humans, animals, or plants. But there are also divergences from those categories in that robots appear resistant to neat and singular classification in any of those categories because they are often perceived in a liminal or "between" space of animate beings and inanimate objects (Kahn Jr. et al., 2011; Kahn Jr. & Shen, 2017). In essence, NOC posits that robots may be perceived as belonging to a new, emergent class. This implies that as robots evolve in design and functionality, our understanding of their place in the world and their potential roles will also evolve because changes in a robot's form can lead to shifts in our understanding of its ontology and role.

In the nearer term, the NOC concept prompts us to reconsider how we interact with and interpret robots in our daily lives, whether in healthcare, education, entertainment, or domestic settings. With each new iteration of robot design and functionality, our understanding of their place in the world and their potential roles can be expected to subtly evolve. This evolution is not driven solely by technological advancements but also by ongoing changes in societal norms, cultural perceptions, and ethical considerations surrounding robots. Moreover, the emergence of robots as a distinct ontological category challenges traditional notions of agency, autonomy, and responsibility (Gunkel, 2023). As robots become more autonomous and capable of independent decision-making, questions arise regarding their moral and legal status. Are robots merely tools to be used by humans, or do they possess inherent rights and responsibilities? How should we ethically design and deploy robots in society to ensure their actions align with human values and norms? By acknowledging the concept of NOC, researchers and policymakers can navigate these complex ethical and societal implications more effectively. By understanding robots as belonging to a potentially unique ontological category, we can develop frameworks and guidelines for their design, deployment, and regulation that prioritize human well-being and ethical considerations while fostering innovation and technological progress. Ultimately, the NOC allows us to shape the future of human-robot interaction in a way that is both responsible and forward-thinking.

Both the FFAB and NOC are concerned with how humans perceive, categorize, and assign roles to robots based on their physical characteristics and attributes. While they approach this topic from different angles, they share common ground in highlighting that our interactions with robots are influenced not only by their technical capabilities but also by our preconceived notions about them based on their appearances. Both acknowledge that human perception is dynamic and evolves with time, experience, and technological changes. Moreover, both emphasize that the relationship between form and function is integral to our perception of robots, and both underline how changes in a robot's appearance can lead to shifts in our understanding of a its ontology and role.

THE SHAPE OF THINGS TO COME: THE PROMISES AND PERILS OF BREAKING THE MOLD

Inspired by the possibility that robots herald the emergence of their own unique ontology, there lies an exciting and relatively underexplored avenue—the design of robots that depart from traditional ontologies, escaping the confines of familiar forms and venturing into uncharted territory of being and becoming. Indeed, the realm of robotics is already evolving beyond the boundaries of familiar humanoids, biomimetic designs, and pre-robotic machines, to explore new frontiers of form and function. When robots break free from traditional molds, they offer both promising possibilities and challenging perils for human-robot interaction.

One of the most intriguing aspects of designing robots beyond known beings is the potential to escape the perceptual baggage, or the constraints and preconceptions tied to familiar associations with other entities. When robots take on unconventional forms, they can transcend the limitations of human or animal anatomy. This departure allows for versatility and specialization, enabling robots to excel in specific tasks and environments where traditional designs may fall short. For instance, robots optimized for space exploration can navigate zero-gravity conditions with precision, revolutionizing our capacity to explore the cosmos. In the field of medicine, specialized robot designs are redefining procedures, particularly in minimally invasive surgeries, enhancing patient outcomes, and reducing the burden on surgeons. These innovations exemplify how unconventional robot designs have already revolutionized entire industries, pushing the boundaries of what is possible.

Perhaps what has received less fanfare is the wonderous space of ontological possibility which opens through morphological innovation. Transformations in the morphology of objects often precipitate profound shifts in their ontological classifications. Two compelling examples from the realms of physics and technology underscore this notion: The transition from a flat Earth model to a spherical one and the evolution of computing technology. In ancient civilizations, the Earth was conceptualized as a flat plane, shaping perceptions of geography and cosmology. However, the realization of the Earth's spherical nature revolutionized our understanding of the cosmos, prompting a reevaluation of humanity's place within it. This shift not only reshaped geographical knowledge and paved the way for contemporary navigational techniques, but it also upended long-standing metaphysical assumptions, challenging established worldviews, and paving the way for modern scientific thought (Kuhn, 2012). Similarly, the evolution of computing devices illustrates how morphological innovations redefine the ontological category of technology (Campbell-Kelly et al., 2023). From the bulky mainframe computers of the past, computers have evolved into sleek and portable devices such as laptops, tablets, and smartphones. This transformation has not only made computing more accessible and integrated into everyday life but has also blurred the very boundaries between work and leisure, public and personal. These examples highlight the profound and sometimes unanticipated implications of morphological innovation, as they challenge established beliefs and reshape our understanding of reality and technology. Such paradigm shifts demonstrate how alterations in physical form can fundamentally reshape ontological and metaphysical assumptions, opening new avenues for exploration and understanding.

Nevertheless, a bold departure from familiar forms also ushers in its own set of challenges. While breaking free from the shackles of familiar associations can be liberating, it concurrently introduces ambiguity, uncertainty, and discomfort into some contexts of human-robot interactions. Therefore, it is vital to recognize that the pursuit of robots beyond traditional frameworks of being is not driven solely by novelty; rather, it encompasses a profound reimagining of what a robot can embody and the multifaceted roles it can assume in our lives. As robots continue to blur ontological boundaries, profound questions about our responsibilities to robots and their reciprocation, the way we ought to perceive them, the identification of our commonalities and distinctions, and the ethical considerations surrounding their use and integration loom larger than ever. These challenges underscore the intricate relationship between form and identity within the context of robotics, where morphology profoundly shapes perceived ontology and societal roles.

REFERENCES

Andriella, A., Torras, C., & Alenyà, G. (2020). Short-term human–robot interaction adaptability in real-world environments. *International Journal of Social Robotics, 12*, 639–657.

Bernotat, J., Eyssel, F., & Sachse, J. (2017). Shape it—The influence of robot body shape on gender perception in robots. In *Proceedings of the International Conference on Social Robotics* (pp. 75–84). Springer.

Blier, S. P. (1994). *The anatomy of architecture: Ontology and metaphor in Batammaliba architectural expression*. University of Chicago Press.

Broussard, M. A., Coates, M., & Martinsen, P. (2023). Artificial pollination technologies: A review. *Agronomy, 13*, 1351.

Campbell-Kelly, M., Aspray, W. F., Yost, J. R., Tinn, H., & Díaz, G. C. (2023). *Computer: A history of the information machine*. Taylor & Francis.

Čapek, K. (2004). *RUR (Rossum's universal robots)*. Penguin. (Original work published 1920)

Cui, J., & Trinkle, J. (2021). Toward next-generation learned robot manipulation. *Science Robotics, 6*, eabd9461.

Edwards, A. (2018). Animals, humans, and machines: Interactive implications of ontological classification. In A. L. Guzman (Ed.), *Human-machine communication: Rethinking communication, technology, and ourselves* (pp. 29–50). Peter Lang.

Edwards, A., Edwards, C., Westerman, D., & Spence, P. R. (2019). Initial expectations, interactions, and beyond with social robots. *Computers in Human Behavior, 90*, 308–314.

Haring, K. S., Watanabe, K., Velonaki, M., Tossell, C. C., & Finomore, V. (2018). FFAB—The form function attribution bias in human–robot interaction. *IEEE Transactions on Cognitive and Developmental Systems, 10*(4), 843–851.

Gunkel, D. J. (2023). *Person, thing, robot: A moral and legal ontology for the 21st century and beyond*. MIT Press.

Hiroi, Y., & Ito, A. (2008). Are bigger robots scary?—The relationship between robot size and psychological threat. In *Proceedings of the International Conference on Advanced Intelligent Mechatronics* (pp. 546–551). IEEE.

Kahn Jr., P. H., Reichert, A. L., Gary, H. E., Kanda, T., Ishiguro, H., Shen, S., Ruckert, J. H., & Gill, B. (2011). The new ontological category hypothesis in human-robot interaction. In *Proceedings of the International Conference on Human-robot Interaction* (pp. 159–160). ACM.

Kahn Jr, P. H., & Shen, S. (2017). NOC NOC, who's there? A new ontological category (NOC) for social robots. In N. Budwig, E. Turiel, & P. D. Zelazo (Eds.), *New perspectives on human development* (pp. 106–122). Cambridge University Press.

Kuhn, T. S. (2012). *The structure of scientific revolutions*. University of Chicago Press.

Kwon, M., Junge, M. F., & Knepper, R. A. (2016). Human expectations of social robots. In *Proceedings of the International Conference on Human-Robot Interaction* (pp. 463–464). IEEE.

Mackenzie, D. (2003). Shape shifters tread a daunting path toward reality. *Science, 301*, 754–756.

Malle, B. F., Fischer, K., Young, J., Moon, A., & Collins, E. (2020). Trust and the discrepancy between expectations and actual capabilities. In D. Zhang and B. Wei (Eds.), *Human-robot interaction: Control, analysis, and design* (pp. 1–23). Cambridge.

Moses, J., & Ford, G. (2021). See Spot save lives: Fear, humanitarianism, and war in the development of robot quadrupeds. *Digital War, 2*(1–3), 64–76.

Palmer, S. E. (1999). *Vision science: Photons to phenomenology*. MIT Press.

Sims, V. K., Chin, M. G., Sushil, D. J., Barber, D. J., Ballion, T., Clark, B. R., Garfield, K. A., Dolezal, M. J., Shumaker, R., & Finkelstein, N. (2005). Anthropomorphism of robotic forms: A response to affordances? In *Proceedings of the Human Factors and Ergonomics Society Annual Meeting, 49*(3), 602–605.

Strait, M. K., Floerke, V. A., Ju, W., Maddox, K., Remedios, J. D., Jung, M. F., & Urry, H. L. (2017). Understanding the uncanny: Both atypical features and category ambiguity provoke aversion toward humanlike robots. *Frontiers in Psychology, 8*, 1366.

Strait, M., Ramos, A. S., Contreras, V., & Garcia, N. (2018). Robots racialized in the likeness of marginalized social identities are subject to greater dehumanization than those racialized as white. In *Proceedings of the International Symposium on Robot and Human Interactive Communication* (pp. 452–457). IEEE.

Taylor, C. (1959). Ontology. *Philosophy, 34*(129), 125–141.

Thomasson, A. L. (2005). The ontology of art and knowledge in aesthetics. *The Journal of Aesthetics and Art Criticism, 63*(3), 221–229.

Weisman, K. (2022). Extraordinary entities: Insights into folk ontology from studies of lay people's beliefs about robots. In *Proceedings of the Annual Meeting of the Cognitive Science Society, 44*(44), 3493–3499. University of California.

Will, P. M., Castaño, A., & Shen, W. M. (1999). Robot modularity for self-reconfiguration. In *Sensor Fusion and Decentralized Control in Robotic Systems II, 3839*, 236–245. SPIE.

CHAPTER TWO

Heads & Faces: An Assemblage of Cues

XIAOYU JIA & CHIEN-HSIUNG CHEN

Humans have long been fascinated by the creation of humanoid robots to accompany or assist us in our daily lives. Initially, researchers developed social robots that mimic human behavior and communication styles to enhance the efficiency and quality of interaction between humans and robots. Due to its initial design focusing on imitating human beings, robots are often designed with a head and face. As early as 1999, Breazeal and Scassellati created a robot named Kismet, which was designed to engage in social interactions with humans using only a head. The robot head was equipped with an active vision system (i.e., it could perform vision-related behaviors for affective expression) and could display a variety of facial expressions. In 2002, DiSalvo and colleagues conducted research on the physical characteristics of social robot faces, finding that certain characteristics, such as the head-size proportions and the total number of facial features significantly influenced human perceptions of the robots' humanity. In 2008, scientists argued that a *social* robot is a robot capable of interacting with people on an interpersonal and socio-emotional level (Breazeal et al., 2008). We contend that the head and face of a robot are central to that social potential. The head and face of a robot can be defined as a distinct part of the robot that is clearly separable from the torso (Strait et al., 2015), where the head is the general segment of the body and the face is the forward-facing portion of the head that generally has features projecting social information, such as eyes or a mouth. To date, robot heads and faces have received a lot of attention because they are intended to be the primary areas for facilitating human-robot interaction (HRI; Breazeal & Scassellati, 1999; Chesher & Andreallo, 2021; DiSalvo et al., 2002; Kalegina et al., 2018; Mathur & Reichling, 2016; Song & Luximon, 2020). It is a "place" that the interaction with humans occurs.

CHARACTERIZING HEADS AND FACES

The mere presence of heads or head-like features on a robot can influence how people perceive it. For instance, merely having a head significantly increases people's willingness to work with it (Kiesler & Goetz, 2002; McGinn, 2020) and may facilitate easier interaction, as it helps focus our interactive efforts on a specific place that is clearly distinct from the less-expressive torso. A humanoid robot's head garners the most visual attention from users, compared to other body parts (Li et al., 2022).

Heads can take many forms, but perhaps their most important function—aside from being a venue for the face—is their movements. The robot's head movements primarily include rotation, tilting, raising, and nodding. These movements depend on the degrees of freedom (DoF) in a robot's neck. DoF specifically refers to the number of axes that the robot (or robot part) can move along. It is one of the key indicators of robot performance, reflecting the robot's mechanical complexity and often its ability to flexibly accomplish tasks, but are also central to the kinematics (or movements) that support social interaction. The head may be able to rotate at multiple angles (like the robot Jibo), but it is also possible that a head may not move at all. The rotation of the robot's head has, to some extent, improved the convenience of communication between humans and robots (Striepe et al., 2021) as head movements perform semantic, narrative, and discursive functions in the process of communication. For example, shaking and nodding is a social behavior of human beings, and a series of movements of robot heads can also be executed based on these human social rules (Poggi & D'Errico, 2011; see Chapter Eight).

The driving mechanisms of the social robot's head and face can be mainly divided into mechanical and digital types (Kalegina et al., 2018; McGinn, 2020). A mechanical social interface utilizes actuators and moving parts to convey social expressions (e.g., Kismet) while a digital interface (e.g., Flobi) achieves this through digital graphics (McGinn, 2020). Furthermore, digital heads can be categorized into three types: Low-resolution LED matrix displays, projected light (reflected against a continuous transparent surface), and high-resolution rendered displays (Chen & Jia, 2023). Faces can also be classified into realistic faces, symbolic faces, blank faces, tech faces, and screens (Chesher & Andreallo, 2021), and they may be expressive or non-expressive

(Rojas-Quintero & Rodríguez-Liñán, 2021). Some faces may have human-like features such as eyes, a nose, and a mouth, while others have none at all (like the robot ASIMO).

Social robots equipped with facial expressions allow humans to identify the robot's expressions of happiness, sadness, anger, disgust, and other simulations of emotions. Humans can also differentiate among varying levels of intensity for the expressed emotions (Kirby et al., 2010), so a robot's facial expressions can convey rich social information that facilitates effective interaction between humans and robots. Creating the face of a robot is a significant challenge because the human face—setting our expectations for facial expressivity—plays a crucial, expressive role in social interaction (Chesher & Andreallo, 2021). It involves many aspects: Physiological, neurological, psychological, cultural, conceptual, affective, esthetic, and ethical. The face of a social robot must be able to express itself as an assembly.

IMPACTS OF HEADS AND FACES ON HRI

In a broad sense, when people interact with robots, they usually talk to or stare at faces or heads (if the robot indeed has them). We tend to treat the robots according to our human rules, and humans tend to apply human social norms to social robots and perceive them as human or creature-like (Nass et al., 1994). We frequently interact with the head of a robot and wait for its responses to ascertain whether it is approaching or moving away from us. For instance, a study found that adopting the intentional stance—when humans infer mindful intentionality to explain and predict other entities' behaviors (Dennett, 1971)—can also occur with respect to social robots (Marchesi et al., 2019). In addition, robots' eye gaze can increase the fluidity of conversation (Mavridis, 2015) or direct a user's attention to relevant information in a tutoring setting (Johnson et al., 2000). The head and face create the impression that a social robot is capable of understanding our instructions and providing responses through facial expressions, head tilting, rotating, or nodding the head, etc. That is, we may hold a "theory of mind" (ToM) in which we attribute beliefs, goals, perceptions, and other mental states to others, allowing individuals to comprehend their behavior and expression within an intentional or goal-oriented framework (Scassellati, 2002). Research indicates that people do sometimes see humanoid robots as having minds, and many of the indicators of that perception are grounded in face-conveyed social information (Banks, 2021).

In that vein, a good deal of evidence suggests the composition of robot heads or faces has a significant impact on the perception of social robots and the outcomes of human-robot interaction. Regarding robot heads, two aspects have meaningful impacts on human perception: Proportions and shape. Firstly, the proportions of a robot's head, such as its width, ratio of eye to face width, and total number of facial features significantly influence the perception of human likeness in robot heads (e.g., DiSalvo et al., 2002; Song & Luximon, 2021; Chen & Jia, 2023). Face size, such as the distance from forehead to chin, can influence a robot's perceived intelligence and sociability (Powers & Kiesler, 2006), while facial proportions are associated with people's perceptions of trustworthiness and their intention to purchase the robot (Song & Luximon, 2021). Combinations of eye and head types prompts different perceptions of robot personality and functionality (Luria et al., 2018).

Secondly, the shape of the robot's head significantly affects the user's perception of the robot across multiple dimensions. A round head was considered more humanlike because it was perceived as being more animated, friendly, intelligent, and feminine than a rectangular head. The shape of a robot's head also impacts human emotional perception of the robot. For example, the more humanlike round head was considered to have greater animacy, friendliness, intelligence, and femininity compared to a rectangular head (Chen & Jia, 2023). Importantly, head shapes often cannot be considered in isolation, and combined with different torsos, and limbs can elicit emotions and influence differential perceptions of a robot's personality. In a study of head, torso, and limb combinations, different head and body combinations provoked different personality evaluations and emotional responses (Hwang et al., 2013). Alongside shape, posture of the head and neck can also be influential: Compared to an upright head posture, identical robots with a tilted head elicited a higher sense of human-likeness, cuteness, and spine-tinglingness (Mara & Appel, 2015).

DiSalvo and his colleagues found that human perception of robot heads is highly dependent on the presence of specific facial features (DiSalvo et al., 2002). The nose, eyelids, and mouth were found to be the facial features that most enhanced a robot's human-likeness. The study also found that when the head was wider than its height, the robot was perceived as more robot-like (and less human-like). Moreover, the appearance became less human-like when the amount of head-surface taken up by the forehead, hair, or chin was reduced (Prakash & Rogers, 2015). Social communication is usually most effectively achieved through the use of local features such as the eyes, mouth, and eyebrows. These features collectively serve as the focal points of interaction, typically regarded as the primary *social interface* for human-robot interaction (McGinn, 2020; see also Chapter Three).

HEADS BEYOND THE HUMANOID?

A humanoid robot is defined as being inspired by the human appearance, structure, and kinematics. So, specifically, a humanoid robot head is supposed to bear some physical and functional resemblance to a human head (Rojas-Quintero & Rodríguez-Liñán, 2021). The presence of humanlike head, face, and body features increases the perceived human-likeness of a robot overall (Erebak & Turgut, 2019). In particular, a social robot's expressions are typically based on the Facial Action Coding System (FACS; e.g., Kędzierski et al., 2013). FACS includes six basic emotions (i.e., happiness, surprise, anger, sadness, fear, and disgust) and was derived from an analysis of the human anatomical basis of facial movement (Ekman & Friesen, 1978). Some scholars have developed highly recognizable and human-like dynamic facial expressions for the six basic emotions (Chen et al., 2018) such that people can correctly perceive the robot's mood and emotion (Fernández-Rodicio et al., 2022) based on facial resemblance to human expressions.

Although a human-like head offers numerous advantages for enhancing HRI (Song & Luximon, 2020), the debate on whether robots should strive to be like humans is ongoing, and the ethical concerns surrounding robots becoming too human-like have gradually garnered attention. From a design standpoint, the human-likeness of robots instigates an uncanny valley effect in which the most mechanical robots elicit nearly neutral reactions, but the likeability of robots increases as its human-resemblance increases up to a point. However, faces became more human than mechanical, they began to be perceived as frankly unlikable, until finally, as faces became nearly human, likability sharply rebounded to a final positive ending point (Mathur & Reichling, 2016). From an ethical standpoint, some ethicists have argued that designing robots to be human-like effectively preys on our fundamentally social nature and constitutes a "superficial state deception" (using a signal to indicate some capacity it lacks) or "hidden state deception" (using a signal to hide a capacity it does have; see Danaher, 2020). Others argue that we are subject to deception in *any* social interaction such that anthropomorphic features should be considered banal (Natale, 2023) especially given that human-like designs—including faces— are necessary to facilitate natural interactions (see Coeckelbergh, 2011).

This debate in mind, the possibility for robot heads and faces to be *non-anthropomorphic* is useful to consider. Specifically, as designed technologies, robots need not necessarily adhere to human-body norms or structures, and can manifest various shapes, structures, and functions for heads and faces—or potentially have none at all.

In science fiction works, people often attribute robots with self-awareness and complex emotions similar to those of real humans, largely showcased through heads. To some extent, they do not even need to be able to speak because they have heads and faces that can express emotions, as seen in characters like Baymax (with its small, ovular head and buttonhole-eyes) and Wall-E (with its binocular-like head that expressively tilts on a triple-jointed neck), in their titular films (Hall & Williams, 2014; Stanton, 2008). Technically speaking, a robot's face can convey expressions in various ways. A head-like feature can express emotions through color, sound, and vibration (Song & Yamada, 2017). For example, Löffler and colleagues (2018) systematically designed and validated a set of 28 different uni- and multi-modal expressions for the basic emotions of joy, sadness, fear, and anger, based on emotion metaphors that capture mental models of emotions. They found that joy was best conveyed through color and motion, sadness through sound, fear through motion, and anger through color. Indeed, people may prefer robot colors that indicate calmness, intelligence, and logic (see Lin et al., 2023). Considering another affordance, a robot's face can be displayed using a digital screen, allowing it to freely show different expressions, animations, or colors—ones that could change based on the audience or situation, rather than being fixed. As can be seen, incorporating colors, motions, and dynamic displays into the design of the human-robot interface could bring it closer to the users' thoughts in ways that reconfigure heads and faces, or perhaps not require a head at all.

It is possible that faceless robots could help avoid some of the discomfort that people may experience around robots. As illustrated in the prior example, the uncanny valley effect manifests when robots with highly humanized appearances fall *just short* of being believable as a healthy human, they can trigger negative emotions in people, ranging from aversion to disgust (Mori et al., 2012). It may be that a robot with no face is preferred over one that provokes uncanny discomfort; however, there may be critical individual and cultural differences. U.S. Americans have shown a significantly stronger preference for robots to have no facial expressions or a face at all compared to Korean and Turkish participants (Lee & Šabanović, 2014), while older adults in Sweden and Italy preferred over a faceless version (Cortellessa et al., 2008). There is yet little research into the effects of *facelessness* as a robot-design paradigm. Beyond this, robots may not necessarily need to have a head that is humanlike in shape but can similarly avoid the uncanny through a more robotic head or a cute head that imitates animals.

Currently, however, social robots are thought to require a head or face to facilitate HRI--or at least a head-like structure. In the future, the human relationship with social robots may be very different from what it is now. Therefore, it is unclear whether the appearance of social robots, particularly their heads and faces, will remain the same in the future. The appearance of social robots will evolve to meet human needs and expectations. In addition, the design of their head or face will play a crucial role in determining the roles they can fulfill in future human society and how they will be perceived and treated by both humans and perhaps by other robots. It is an open question as to whether robots will still require a head or face—or if some other *inter*face may suffice.

REFERENCES

Banks, J. (2021). Of like mind: The (mostly) similar mentalizing of robots and humans. *Technology, Mind, and Behavior, 1*(2).

Breazeal, C., & Scassellati, B. (1999). A context-dependent attention system for a social robot. In *Proceedings of the International Joint Conference on Artificial Intelligence* (pp. 1146–1153). IJCAI.

Breazeal, C., Takanishi, A., & Kobayashi, T. (2008). Social robots that interact with people. *Springer handbook of robotics* (pp. 1349–1369). Springer.

Chen, C. H., & Jia, X. (2023). Effects of head shape, facial features, camera, and gender on the perceptions of rendered robot faces. *International Journal of Social Robotics, 15*(1), 71–84.

Chen, C., Garrod, O. G., Zhan, J., Beskow, J., Schyns, P. G., & Jack, R. E. (2018, May). Reverse engineering psychologically valid facial expressions of emotion into social robots. In *Proceedings of the International Conference on Automatic Face & Gesture Recognition* (pp. 448–452). IEEE.

Chesher, C., & Andreallo, F. (2021). Robotic faciality: The philosophy, science and art of robot faces. *International Journal of Social Robotics, 13*(1), 83–96.

Coeckelbergh, M. (2011). Are emotional robots deceptive? *IEEE Transactions on Affective Computing, 3*(4), 388–393.

Cortellessa, G., Scopelliti, M., Tiberio, L., Svedberg, G. K., Loutfi, A., & Pecora, F. (2008, November). A cross-cultural evaluation of domestic assistive robots. In *Proceedings of the AAAI Fall Symposium: AI in Eldercare: New Solutions to Old Problems* (pp. 24–31). AAAI.

Danaher, J. (2020). Robot betrayal: A guide to the ethics of robot deception. *Ethics and Information Technology, 22*, 117–128.

Dennett, D. C. (1971). Intentional systems. *The Journal of Philosophy, 68*(4), 87–106.

DiSalvo, C. F., Gemperle, F., Forlizzi, J., & Kiesler, S. (2002). All robots are not created equal: The design and perception of humanoid robot heads. In *Proceedings of the Conference on Designing Interactive Systems: Processes, Practices, Methods, and Techniques* (pp. 321–326). ACM.

Ekman, P., & Friesen, W. V. (1978). *Facial action coding system.* Consulting Psychologists Press.

Erebak, S., & Turgut, T. (2019). Caregivers' attitudes toward potential robot coworkers in elder care. *Cognition, Technology and Work, 21*, 327–336.

Fernández-Rodicio, E., Maroto-Gómez, M., Castro-González, Á., Malfaz, M., & Salichs, M. Á. (2022). Emotion and mood blending in embodied artificial agents: Expressing affective states in the mini social robot. *International Journal of Social Robotics, 14*(8), 1841–1864.

Hall, D., & Williams, C. (2014). *Big Hero 6* [Film]. Walt Disney Animation Studios.

Hwang, J., Park, T., & Hwang, W. (2013). The effects of overall robot shape on the emotions invoked in users and the perceived personalities of robot. *Applied Ergonomics, 44*(3), 459–471.

Johnson, W. L., Rickel, J. W., & Lester, J. C. (2000). Animated pedagogical agents: Face-to-face interaction in interactive learning environments. *International Journal of Artificial Intelligence in Education, 11*(1), 47–78.

Kalegina, A., Schroeder, G., Allchin, A., Berlin, K., & Cakmak, M. (2018, February). Characterizing the design space of rendered robot faces. In *Proceedings of the International Conference on Human-Robot Interaction* (pp. 96–104). ACM.

Kędzierski, J., Muszyński, R., Zoll, C., Oleksy, A., & Frontkiewicz, M. (2013). EMYS—Emotive head of a social robot. *International Journal of Social Robotics, 5*, 237–249.

Kiesler, S., & Goetz, J. (2002). Mental models and cooperation with robotic assistants. In *Proceedings of the Conference on Human Factors in Computing Systems* (pp. 576–577). ACM.

Kirby, R., Forlizzi, J., & Simmons, R. (2010). Affective social robots. *Robotics and Autonomous Systems, 58*(3), 322–332.

Lee, H. R., & Šabanović, S. (2014, March). Culturally variable preferences for robot design and use in South Korea, Turkey, and the United States. In *Proceedings of the International Conference on Human-Robot Interaction* (pp. 17–24). ACM.

Li, M., Guo, F., Ren, Z., & Duffy, V. G. (2022). A visual and neural evaluation of the affective impression on humanoid robot appearances in free viewing. *International Journal of Industrial Ergonomics, 88*, 103159.

Lin, P. C., Hung, P. C., Jiang, Y., Velasco, C. P., & Martínez Cano, M. A. (2023). An experimental design for facial and color emotion expression of a social robot. *The Journal of Supercomputing, 79*(2), 1980–2009.

Löffler, D., Schmidt, N., & Tscharn, R. (2018). Multimodal expression of artificial emotion in social robots using color, motion and sound. *Proceedings of the International Conference on Human-Robot Interaction* (pp. 334–343). ACM.

Luria, M., Forlizzi, J., & Hodgins, J. (2018, August). The effects of eye design on the perception of social robots. In *Proceedings of the International Symposium on Robot and Human Interactive Communication* (pp. 1032–1037). IEEE.

Mara, M., & Appel, M. (2015). Effects of lateral head tilt on user perceptions of humanoid and android robots. *Computers in Human Behavior, 44*, 326–334.

Marchesi, S., Ghiglino, D., Ciardo, F., Perez-Osorio, J., Baykara, E., & Wykowska, A. (2019). Do we adopt the intentional stance toward humanoid robots? *Frontiers in Psychology, 10*, 450.

Mathur, M. B., & Reichling, D. B. (2016). Navigating a social world with robot partners: A quantitative cartography of the Uncanny Valley. *Cognition, 146*, 22–32.

Mavridis, N. (2015). A review of verbal and non-verbal human–robot interactive communication. *Robotics and Autonomous Systems, 63*, 22–35.

McGinn, C. (2020). Why do robots need a head? The role of social interfaces on service robots. *International Journal of Social Robotics, 12*(1), 281–295.

Mori, M., MacDorman, K. F., & Kageki, N. (2012). The uncanny valley [from the field]. *IEEE Robotics & Automation Magazine, 19*(2), 98–100.

Nass, C., Steuer, J., Henriksen, L., & Dryer, D. C. (1994). Machines, social attributions, and ethopoeia: Performance assessments of computers subsequent to "self-" or "other-" evaluations. *International Journal of Human-Computer Studies, 40*(3), 543–559.

Natale, S. (2023). AI, human-machine communication and deception. In A. Guzman, R. McEwen, & S. Jones (Eds.), *The SAGE handbook of human-machine communication* (pp. 401–408). SAGE.

Poggi, I., & D'Errico, F. (2011). Social signals: A psychological perspective. In *Computer analysis of human behavior* (pp. 185–225). London: Springer.

Powers, A., & Kiesler, S. (2006, March). The advisor robot: Tracing people's mental model from a robot's physical attributes. In *Proceedings of the International Conference on Human-Robot Interaction* (pp. 218–225). ACM.

Prakash, A, & Rogers W. A. (2015). Why some humanoid faces are perceived more positively than others: Effects of human-likeness and task. *International Journal of Social Robotics, 7*(2), 309–331.

Rojas-Quintero, J. A., & Rodríguez-Liñán, M. C. (2021). A literature review of sensor heads for humanoid robots. *Robotics and Autonomous Systems, 143*, 103834.

Scassellati, B. (2002). Theory of mind for a humanoid robot. *Autonomous Robots, 12*, 13–24.

Song, S., & Yamada, S. (2017, March). Expressing emotions through color, sound, and vibration with an appearance-constrained social robot. In *Proceedings of the International Conference on Human-Robot Interaction* (pp. 2–11). ACM.

Song, Y., & Luximon, Y. (2020). Trust in AI agent: A systematic review of facial anthropomorphic trustworthiness for social robot design. *Sensors, 20*(18), 5087.

Song, Y., & Luximon, Y. (2021). The face of trust: The effect of robot face ratio on consumer preference. *Computers in Human Behavior, 116*, 106620.

Stanton, A. (2008). *WALL-E* [Film]. Pixar Animation Studios.

Strait, M., Vujovic, L., Floerke, V., Scheutz, M., & Urry, H. (2015, April). Too much humanness for human-robot interaction: Exposure to highly humanlike robots elicits aversive responding in observers. In *Proceedings of the Conference on Human Factors in Computing Systems* (pp. 3593–3602). ACM.

Striepe, H., Donnermann, M., Lein, M., & Lugrin, B. (2021). Modeling and evaluating emotion, contextual head movement and voices for a social robot storyteller. *International Journal of Social Robotics, 13*, 441–457.

CHAPTER THREE

Eyes & Gaze: Exchanging Glances with Robots

CHRIS CHESHER

Walking into the Pepper Parlor concept café in Tokyu Plaza Shibuya, Tokyo, it is easy to catch the eye of one of 30 or so white, humanoid Pepper robots that inhabit this space. I move in front of one and it locks on to my face with its gently blinking eyes, following me as I move. This positions me socially and physically in relation to the robot, giving me implicit permission to interact and aligning me with the microphones and touchscreen. Having captured my gaze through these non-verbal devices, it holds me in its gaze, fidgeting, flexing its hands and urging me on its screen: "Please talk."

The initially peculiar sense that Pepper has met my gaze recalls familiar and sometimes transformative experiences of sharing other people's gazes. Human eyes meet in a variety of social situations often with meaning and feeling: Intimacy, menace, ridicule, or awkwardness. But I know that this is a robot, and not another human, and this leaves me perplexed. Perhaps this is simply a surface effect: A programmed visual mimicking of eye contact with nothing behind it, nudging me automatically into verbal interaction. But maybe it is more, as the eye contact could indicate I am being analyzed by machine intelligence or some remote human operator. Pepper's fixed, wide-eyed expression gives little away. Walking into the Pepper Parlor to find a table, I'm surrounded by these gazing non-humans making me feel I'm in the presence of a crowd of others witnessing my every move.

EYES AND THE INTERACTIVE GAZE

Biological eyes are the second-most complex organ in animals and humans, after the brain. Of course, they are central to the sense of sight. But they are also essential in interacting with other people. Blinking and eye movements indicate someone is alive. The shape, color, and position of people's eyes on their faces help others recognize them as unique individuals. The direction in which a person's head and eyes are looking communicates social and spatial information about what they are paying attention to—their gaze.

Like Pepper, many robots are designed to use their eyes and gaze in interacting with people. Pepper's eyes glow in different colors and blink occasionally. Its childlike, cute, and oversized eyes above an abstracted nose and mouth make it distinctive and friendly-looking, even though it cannot change its expression. Pepper's eyes are not capable of vision. Instead, it has two high-definition color cameras—one in its forehead and another in its mouth—as well as depth sensors behind its eyes. These sensors allow Pepper's software to analyze objects, detect and recognize faces, and analyze gaze direction. Using this data, Pepper can interact physically and socially with its environment. As social robotics has developed, recognizing and performing gazes has proved among the more useful functions in human-robot interaction: A nonverbal mode of communication that can be as nuanced as it is powerful.

BEYOND BIOLOGY: THE AESTHETICS AND POLITICS OF THE GAZE

Although robot eyes and vision are often compared to biological equivalents, they are more usefully understood as cultural artefacts and media technologies. First, robot cameras and sensors are technical objects with different sensitivities, affordances, and costs (see Chapter Twelve). They are developed by engineering companies, selected by robot developers, and then programmed to interact with people and the world. Second, gazes are familiar from art, communicating emotions, representing social relationships, and telling stories. Gazes are central to cinema's capacity to engage viewers with the physical and social world of a film. The camera's gaze moves and shifts perspective with each edit. The characters' gazes are key to their dramatic interaction. Meanwhile, cinema spectators find themselves gazing with the camera's perspective at the action on the screen. Mulvey (1975) argues that mainstream cinema often favors the unconscious pleasure of male characters and viewers. But unlike cinema, robot gazes meet peoples' eyes, seemingly subjecting them to visual interrogation and demands.

Therefore, robot design works with a repertoire of technical and aesthetic elements that allow the robot to technically, artfully, and interactively perform gazes and to be gazed upon. Depending on the robot, movements of its head, eyes, and body can communicate gazes with audiences. Gaze can communicate the robot's active status, its locus of attention, its ostensible emotional state, and its future intentions. But what about the question of identity in human-robot gazes? Whenever a person's gaze encounters another person's gaze, the identities of participants are crucial. Are they strangers, lovers, judge and criminal, worker and boss? What is their gender, class, and race? In Pepper Parlor, some people may recognize Pepper as a celebrity robot model, but there is no individual identity for any of the robots. Each is interchangeable with any of the others. Their snow-white skin and androgenous appearances also leave audiences uncertain about their gender and race (Sparrow, 2020). The robots perform as anonymous service workers—hosts and entertainers. Each is assigned to a table where it can attempt to establish a relationship with one group of diners.

Recognition can theoretically work in the other direction—the robot might recognize visitors. Whether or not face recognition is used in this café, the Pepper robot is certainly designed with a capacity to recognize people, discriminate among them, and record their actions. In this way, Pepper fits within a historical lineage of technologized and institutionalized power of the disciplinary gaze. For example, in prisons, schools, and hospitals, certain subjects have the power to gaze. Prison guards, teachers, and doctors may gaze while prisoners, students, and patients are arranged so they can be gazed upon (Foucault, 1995). Pepper is designed to both gaze and be gazed upon. But other robots in institutions such as shopping centers, police beats, and military battlefields are able to perform surveillance more effectively. Therefore, a cultural and media theory of robotic gazes should analyze these dimensions of technology—art, identity, and institutionalization. It should also analyze the meanings of the gazes we come to in the next section—the assertiveness of direct gazes, the submissiveness of averted gazes, and the orientations in space of deictic gazes.

LOOK AT ME: DIRECT GAZE

The experience of having eye contact with Pepper exercises one of the most powerful modes of non-verbal interpersonal communication: Direct gaze, when someone looks directly into the eyes of another, focusing their attention toward them. Classic social psychology research in the 1970s showed that direct eye contact can carry quite different meanings. If it is accompanied by angry facial contortions and verbal abuse, a direct gaze contributes to expressions of aggression

(Ellsworth, 1975). Even in less intense encounters, unconscious negotiations of eye contact can establish interpersonal power. The dominant person sustains eye contact, while the "weaker" person averts their gaze in submission (Ellsworth & Carlsmith, 1973). On the other hand, in situations of intimacy, direct gaze can contribute to a mutual sense of attraction, closeness, and trust. When gaze is shared by two parties for a sustained period, it is considered mutual gaze (Argyle & Cook, 1976), which can help build intimacy.

The direct gaze of a subject toward its viewers is often found in traditional media, allowing them to construe meanings and experience emotions. In the famous First World War recruiting image headlined with the statement "Your country needs you," the insistent gaze of Field-Marshal Lord Kitchener directly addresses the viewer to command immediate attention. Semioticians Kress and Van Leeuwen (2006) use it as example of "direct address" (p. 117), where Kitchener's eyes look directly at the viewer, making a "demand" (p. 118; see Fig. 03.1). The face and foreshortened hand's pointing gesture perform an "image act" (p. 117) with great force at this historical moment. The meanings of the poster are formed through a combination of the universal interpersonal experience of direct gaze and the written and visual codes used in the poster.

The gazes of the robots in Pepper Parlor also make demands on people passing by, capturing their attention and inviting them to enter the restaurant. Once

Figure 03.1: A 1914 British military recruitment advertisement depicting British Secretary of State for War (left; source: public domain), and the robot "Pepper" by SoftBank (right; source: Autumn Edwards).

Pepper has a visitor's attention, it has capacities a recruiting poster does not: To automatically identify, analyze, and follow viewers using face recognition, emotion detection, and gaze tracking. It can shift its gaze based on the direction of sound and on movement sensed by microphones, cameras, and a depth sensor. While these functions are not necessarily evident to visitors dining at Pepper Parlor, they represent early steps in developing gaze interactions between people and robots that might apply in contexts where even more is at stake: in advertising, security, policing and even the military.

LOOKING AWAY: AVERTED GAZES

Direct gaze can often lead to discomfort; thus, in everyday interactions, individuals often break eye contact after a brief period. This practice of gaze aversion is interpersonally meaningful, frequently being driven by a desire to modulate intimacy levels. Gaze aversion has a variety of possible meanings, including shyness, modesty, flirtatiousness, submissiveness, or even dishonesty. In some cases, individuals look away simply to give them time to process their thoughts (Doherty-Sneddon et al., 2005).

Gaze relations are also important in the visual grammar of images. In photographs, there is a significant difference between images of people looking at the camera and images in which people are looking down or looking away. Kress and van Leeuwen (2006) describe the depiction of indirect gaze as making the viewer an "offer" (p. 119) rather than a demand. In this case, the connection with the subject is more indirect and the viewer is transformed into an invisible onlooker. The image of the other is presented as an object for dispassionate examination without making any direct contact. In many contexts, photographs depict people with less social power looking away from the camera, implying that the viewer does not need to form a connection with them.

Pepper's default interactive gaze-seeking avoids having the robot seem shy or submissive. Submission may draw attention to the robot's material existence as an electronic gadget that can be examined dispassionately. Even without moving facial features, Pepper's wide-eyed expression and direct gaze evokes optimism and friendliness but sometimes comes across as a little intense. With programming, gaze aversion could soften the intensity of eye contact, potentially allowing conversations to flow more naturally because people subconsciously interpret a robot's gaze aversion as deliberate (Andrist et al., 2014).

LOOKING ELSEWHERE: DEICTIC GAZES

Another context where people look away is when they are redirecting their attention to events elsewhere in the visual world. This is known as deictic gaze and communicates spatial information to those who witness it. It often triggers the phenomenon of gaze cueing: When we see someone shift their gaze to focus on something in the surroundings, we are compelled to look in the same direction. Following gazes gives viewers a sophisticated and automatic understanding of what others are perceiving and what they are interested in (Shepherd, 2010). As following another's gaze involves imagining what they are seeing and paying attention to, this phenomenon is associated with the development of a theory of other peoples' minds (see Chapter Twenty Five).

Another way to understand the deictic gaze is to look at photographs. For Kress and van Leeuwen (2006), images can be analyzed for gaze vectors—imaginary lines emanating from subjects' eyes—that can be followed through the space of the image, which sometimes constitute a visual narrative. For example, the famous "distracted boyfriend" meme has several vectors that combine to tell a story. There is an implied vector indicated by the couples' bodies' orientation as they walk down the street. Another vector is drawn when the man turns around to ogle a woman walking in the other direction. There is a third vector when the woman who is presumably his girlfriend looks at him in horror. Creators of memes based on this image exploit the vectoral narrative diagram, changing the characters in the narrative by adding labels to each of them and making a new meaning based on the existing relationships. Recent vision technologies are now analyzing and diagramming such gaze vectors in images (Kellnhofer et al., 2019) and similar technology allows robots to estimate gaze direction (Palinko et al., 2016).

Robots' heads and eyes can themselves perform gaze vectors that communicate implied perception and intentions. In interactive mode, Pepper can shift its head in response to a nearby sound or movement, creating its own deictic gaze vectors. However, its repertoire of gaze actions in this mode is limited. If developers want to control Pepper's gaze more directly, they can use the Choregraphe software suite to script Pepper's movements along a timeline from pose to pose as in traditional 2D animation. For example, in Pepper Parlor, visitors can instruct Pepper to dance to "Gangnam Style" and other songs. In response, Pepper's body and gaze move rhythmically as it follows the dance moves, but the animated performance lacks the emergent interactivity of unscripted interaction. For Pepper to perform more complex moves, developers use software to program gaze interactions.

Creating effective human-robot interactions is a nuanced art that allows audiences to follow a robot's gaze cues more seamlessly. For example, Chien et al. (2022) programmed Pepper to assist participants in an online shopping activity.

Pepper could perform anticipatory actions such as making eye contact with the shopper and, in the process, actively help them narrow down their selections. Pepper would occasionally tilt its head forward to gaze deictically at the screen on its chest where the choices were displayed. The researchers found that in comparison to those who interacted with Pepper in a more reactive and impassive mode, those who experienced more proactive and expressive behaviors reported a greater sense of intimacy with the robot and trusted it more, both cognitively and affectively.

Even with sophisticated programming, the effectiveness of robot gazes on human perception and behavior remains controversial, but much research indicates that reading robot gaze is a learned skill. For example, Morillo-Mendez et al. (2023) found that observers must form the belief that a robot's gaze is communicating its intentions, allowing them to attribute mental states to the robot. By contrast, Friebe et al. (2022) found that participants responded to gaze cues even when a robot appeared on-screen. Marchesi et al. (2023) found cultural differences in how participants responded to a robot's gaze cues, arguing that Singaporean participants had a higher degree of "cultural attunement" to robots than Italian participants (p. 9). Belkaid et al. (2021) discovered that individuals take longer to respond to prompts when they are engaged in eye contact with a robot as opposed to when the robot is looking away, probably because of the increased cognitive load experienced during mutual gaze. Further studies have found that individuals need sustained interaction with a robot before being able to follow its gaze with confidence (Admoni & Scassellati, 2017).

Another robot capable of emotion and storytelling with gaze behaviors is Engineered Arts' Ameca. The short video "Ameca Humanoid Robot AI Platform" released on YouTube (Engineered Arts, 2021) introduced it to the world with an animated performance that makes heavy use of shifting gaze vectors. When the video opens, Ameca is in a research lab with its eyes closed. When its eyes open, a narrative of visual discovery unfolds. It looks around as if startled by the world, discovering its own hands and examining them closely. It then reels backwards when it spots something in the distance, and finally settles its gaze on the viewer, with a smile. Scripted interactions like this can use gaze shifts to communicate meanings, narratives, and emotions. Ameca's animatronic eyes, face, and head are far more complex than Pepper's, allowing its gaze behaviors to be more naturalistically expressive.

During the interactive operation of both Pepper and Ameca, an array of software-driven sensors continuously measures and evaluates the immediate world of subjects and objects (see Chapter Twelve). They can attempt to recognize the faces of multiple people, estimate their ages, identify emotions, and determine peoples' bodily and head orientations. For example, the Visage software controlling Ameca allows it to perform "real-time tracking of 3D head pose, gaze direction, and facial feature coordinates" (Engineered Arts, 2018, para. 5). These advanced surveillance functions support more effective human-robot interaction, potentially customized for use in locations such as museums, retail outlets, or airports.

As mentioned above, human gaze interactions are always conditioned by the relationships between participants with distinct identities and positions of social power. Research on human interaction suggests that people of different genders (Heru, 2003) and races (Weisbuch et al., 2017) use gaze differently. It is interesting, therefore, that the designers of Pepper and Ameca chose to make the robots' gender, race, and age obscure. The robots' identities are open to interpretation. Ameca's features are androgenous, its skin grey, and its head bald, rendering its gender and race indeterminate. In social semiotic terms, Ameca and Pepper are caught between being subjects and objects—gazing or being gazed upon, even if both are engineered to perform mutual gaze with people nearby. With ambiguous identities, these robots are semiotically interpretable, allowing those who exhibit them to exploit their flexible identities with programming and contextualizing it in a physical, social, and narrative context such as an art gallery, airport, lab, or shopping center.

A more hard-edged technology is the Knightscope K5 droid-shaped security robot, which is aimed at deterring crime with its intimidating appearance (being gazed at) and witnessing crime with its array of sensors (gazing). It has no eyes or face, but a technologized gaze with four prominent cameras positioned around its curved conical body to survey 360 degrees of its surroundings. It can patrol a defined territory such as a parking lot, shopping center, or public park, recognizing faces and numberplates and recording large amounts of data. Being caught in such a robot's gaze may make you aware of its disciplinary power. While it might make some people feel safer, it can also engender a sense being watched and even shamed. In this way, it recalls the logic of panopticism (Foucault, 1995) in institutions such as schools, prisons, hospitals, and workplaces. Within these architectures, occupants are monitored, normalized, and punished. By contrast, security robots are intelligent, mobile incarnations of everyday surveillance familiar from security cameras and other sensors. But unlike these fixed devices, the robot can patrol public spaces, witness events, identify suspects, and even pursue them through remote operation. As we find ourselves increasingly subject to the technologies of the machine gaze, manifested through AI-equipped robots, we become acutely aware of our vulnerability to automatic detection and reporting systems—systems designed to identify individuals exhibiting abnormal behavior or gaze patterns.

CONCLUSIONS

The technologized gaze provides robots with a powerful repertoire of actions that create sensations, meanings, stories, and feelings. As we have seen, robots are developing more sophisticated capacities to perform gazes: Making eye contact with people to make interpersonal demands, bearing witness to their audience's behavior, responding to what they see, displaying it to remote operators, and recording it to an archive. The power of the gaze is a precursor to the physical and social actions robots perform: Interpreting and performing gestures, movements, speech, and so on. But the power of robot gazes to attract, deter, build intimacy, or persuade raises ethical and regulatory questions about robots' capacities for social manipulation, intrusion on privacy and misdirection. These concerns will only become more urgent as technology develops and finds applications that enhance the power of robots' masters.

REFERENCES

Admoni, H., & Scassellati, B. (2017). Social eye gaze in human-robot interaction: A review. *Journal of Human-Robot Interaction, 6*(1), 25–63.

Andrist, S. Tan, X. Z., Gleicher, M., & Mutlu, B. (2014). Conversational gaze aversion for human-like robots. *ACM/IEEE International Conference on Human-Robot Interaction*, 25–32.

Argyle, M., & Cook, M. (1976). *Gaze and mutual gaze*. Cambridge University Press.

Belkaid, M., Kompatsiari, K., Tommaso, D. De, Zablith, I., & Wykowska, A. (2021). Mutual gaze with a robot affects human neural activity and delays decision-making processes. *Science Robotics, 6*(58).

Chien, S. Y., Lin, Y. L., & Chang, B. F. (2022). The effects of intimacy and proactivity on trust in human-humanoid robot interaction. *Information Systems Frontiers, 26*(1), 75–90.

Doherty-Sneddon, G., & Phelps, F. G. (2005). Gaze aversion: A response to cognitive or social difficulty? *Memory and Cognition, 33*(4), 727–733.

Ellsworth, P. C. (1975). Direct gaze as a social stimulus: The example of aggression. In *Nonverbal communication of aggression* (pp. 53-75). Boston, MA: Springer US.

Ellsworth, P., & Carlsmith, J. M. (1973). Eye contact and gaze aversion in an aggressive encounter. *Journal of Personality and Social Psychology, 28*(2), 280–292.

Engineered Arts. (2018). *Sensors* [wiki entry]. <https://wiki.engineeredarts.co.uk/Sensors>

Engineered Arts. (2021). *Ameca Humanoid Robot AI Platform* [Video]. YouTube. <https://youtu.be/IPukuYb9xWw?si=ykKbC1Ip2Nw8Ks8F>

Foucault, M. J. (1995). *Discipline and punish*. Vintage Books.

Friebe, K., Samporová, S., Malinovská, K., & Hoffmann, M. (2022). Gaze cueing and the role of presence in human-robot interaction. In *Proceedings of the International Conference on Social Robotics* (pp. 402–414). Springer Nature Switzerland.

Heru, A. M. (2003). Gender and the gaze: A cultural and psychological review. *International Journal of Psychotherapy, 8*(2), 109–116.

Kellnhofer, P., Recasens, A., Stent, S., Matusik, W., & Torralba, A. (2019). Gaze360: Physically unconstrained gaze estimation in the wild. *Proceedings of the International Conference on Computer Vision* (pp. 6911–6920). IEEE.

Kress, G., & Leeuwen, T. van. (2006). *Reading images. The grammar of visual design* (Second Edition). Taylor & Francis.

Marchesi, S., Abubshait, A., Kompatsiari, K., Wu, Y., & Wykowska, A. (2023). Cultural differences in joint attention and engagement in mutual gaze with a robot face. *Scientific Reports, 13*(1), 1–12.

Morillo-Mendez, L., Stower, R., Sleat, A., Schreiter, T., Leite, I., Mozos, O. M., & Schrooten, M. G. S. (2023). Can the robot "see" what I see? Robot gaze drives attention depending on mental state attribution. *Frontiers in Psychology, 14*, 1215771.

Mulvey, L. (1975). Visual pleasure and narrative cinema. *Screen, 16*(3), 6–18.

Palinko, O., Rea, F., Sandini, G., & Sciutti, A. (2016, October). Robot reading human gaze: Why eye tracking is better than head tracking for human-robot collaboration. In *Proceedings of the International Conference on Intelligent Robots and Systems* (pp. 5048–5054). IEEE.

Shepherd, S. V. (2010). Following gaze: Gaze-following behavior as a window into social cognition. *Frontiers in Integrative Neuroscience*, 4, 1–13.

Sparrow, R. (2020). Robotics has a race problem. *Science, Technology, & Human Values, 45*(3), 538–560.

Weisbuch, M., Pauker, K., Adams, R. B., Lamer, S. A., & Ambady, N. (2017). Race, power, and reflexive gaze following. *Social Cognition, 35*(6), 619–638.

CHAPTER FOUR

Legs & Feet: Complexities of Limb-Environment Interactions

ZHENYU GAN

In the natural world, the ability to move effectively is essential for animals as they search for food and resources. Over time, many species have developed limbs and feet, allowing them to overcome gravity's pull and move across various challenging landscapes. This fascinating development in nature has inspired humans to create robots with similar features. Engineers and researchers, drawing inspiration from the animal kingdom, have designed robots with legs and feet, mimicking the way humans and other animals move. The evolution of legged robots, while not always the most efficient mode of locomotion, emerges as the most adaptive and versatile when operating in human-inhabited spaces or traversing wild areas devoid of paved paths. Unlike robots that roll on wheels or fly, legged robots can navigate through human-centric spaces with ease. They can climb stairs, step over objects, and move around in cluttered spaces. Their capabilities extend well beyond indoor environments. These robots are adept at navigating challenging outdoor terrains as well, from sandy beaches and muddy trails to rocky hillsides. Additionally, they can skillfully maneuver through intricate areas brimming with narrow passages, dense underbrush, or overhanging tree branches. They can assist in industrial workplaces, help in search and rescue operations where navigating difficult terrains is crucial, and even be used in exploring outer space.

In the field of legged robotics, the creation of legs and feet goes far beyond simply replicating human or animal anatomy. It requires an in-depth grasp of physics and material science, coupled with advanced knowledge in the principles of robotics control and environment perception. Engineers and computer scientists apply this knowledge to create structures capable of adapting to various terrains, maintaining dynamic balance, and interacting intelligently with their surroundings. Creating robotic legs and feet is a true blend of technical skill and creative thinking, aiming to make robots that are practical and efficient.

SHAPES OF ROBOT LEGS AND FEET

While it may appear that the concept of legged robots is a recent development, the origins of legged locomotion in robotics date back several centuries. Leonardo da Vinci, between 1495 and 1497, envisioned and possibly constructed the first articulated anthropomorphic robot. In the mid-19th century, Russian mathematician Pafnuty Chebyshev presented a locomotion model utilizing mechanical linkages for horizontal movement while enabling feet to support weight alternately during stepping (Raibert, 1986). A significant leap was Georges Moore's creation of "The Steam Man" in 1893, a bipedal machine powered by a 0.5 horsepower gas-fired boiler, achieving speeds of up to 14 kilometers per hour (Silva & Machado, 2006). Throughout this era, most walking machines relied on kinematic linkages to facilitate movement. In tandem, the mechanical design of robot legs often imitated the joint structures of legged animals, with most robots like RABBIT (Westervelt et al., 2018) and Honda's ASIMO featuring forward-bending knees (Kajita & Espiau, 2008). Others, such as Cassie, drew inspiration from avian leg structures, with knee-like tarsus joints bending backward (Gong et al., 2019). Researchers discovered that these revolute joints, which enable two rigid bodies to rotate around a shared axis, were crucial for creating ground clearance and for energy distribution or power generation during movement.

In the 1980s, a departure from nature-inspired designs emerged. Marc Raibert, founder of Boston Dynamics, developed hopping robots with telescoping, pogo stick-like legs, demonstrating that revolute joints were not necessary for effective movement. He found a nearly linear relationship between forward speed and the landing angle of the leg, leading to self-stabilization and dynamic movements like backflips (Raibert, 1986). Another innovative design came from Oregon State University's Dynamic Robotics Laboratory with the ATRIAS Biped Walking Robot (Hubicki et al., 2018). Its unique parallelogram leg linkages allowed heavy actuators and transmission systems to be mounted near the torso, reducing leg weight and rotational inertia. This design gained popularity for its agility and the potential simplification of motion-control calculations. In essence, no single leg design emerges as universally superior; the effectiveness of a design largely depends on the robot's ability to adjust its leg angle (the angle from the hip to the point of foot contact) and leg length (the distance from the hip to the foot contact) flexibly during both stance and swing phases. With these capabilities, a robot can achieve stability and mobility across various environments.

Engineers and scientists have also delved into a wide range of foot designs for robots, addressing the complexities of foot-environment interactions and collisions. Broadly, these designs can be categorized into three types: Point contact feet, line contact feet, and surface contact feet. The point contact foot, often the simplest to model and design, is exemplified in early designs like the StarlETH quadrupedal robot, which used Racquetballs to minimize rebound upon ground contact (Hutter et al., 2012). This type of foot is particularly adept at navigating uneven terrains, as it doesn't require selective footholds. However, it only provides a force in certain directions at the contact point, making balance challenging for robots with fewer than three legs on the ground simultaneously. Robots with point contact feet resemble humans walking on stilts, constantly moving to maintain equilibrium. Similar to tables with four legs, quadrupedal robots often employ this design, as their four legs provide sufficient actuation and forces to stand still.

For bipedal stability, designs like Cassie feature narrow, rubber feet with line contact, enhanced by additional motors for steadiness. This design not only offers greater friction to prevent foot twisting during the stance phase but also allows bipedal robots to stand still on both feet, although balancing on a single leg can be precarious. Most bipedal robots, however, utilize flat feet with surface contact, requiring dual actuations for pitching and rolling motions. While this design enables balancing on a single leg when the upper body is properly positioned, navigating uneven terrain poses significant challenges. Selecting appropriate footholds and ensuring full contact between foot and ground are critical to prevent slippage and falls. Lastly, inspired by human locomotion—where the heel strikes first and then rolls towards the toe for propulsion—some robotic and prosthetic foot designs, like RHex (Spagna et al., 2007) and "running blades" mimic this rolling cylinder motion. Extensive studies have shown that this heel-to-toe transition increases stability and step length during walking, demonstrating its effectiveness in robotic applications.

CONTROLS OF LEGGED MOTION

Controlling the movement of robots, especially those that walk on two or more legs, is a fascinating yet challenging task. One of the key goals is to keep the robot balanced while it moves. Think of it like trying to keep a pencil balanced on your finger—it requires constant adjustments to your hand and finger in relation to the tipping and swaying of the pencil. In the 1980s, researchers at Waseda University in Japan developed a stable walking controller for a robot called WL-10RD. They discovered that by carefully controlling the robot's center of gravity within the area supported by its feet (imagine an invisible area drawn around the robot's feet), they could prevent it from falling (Sakagami et al., 2002). This idea was later adopted by famous humanoid robots like Honda's ASIMO. Around the same time, Marc Raibert at MIT's Legged Lab was working on robots with one, two, or four legs. He found that by adjusting the angle at which the robot's feet hit the ground and controlling the force with which they pushed off, he could make the robots move dynamically (Raibert, 1986). This method allowed for more agile movements like trotting and was used in later robots such as Boston Dynamics' BigDog.

In the early 2000s, a new concept called Hybrid Zero Dynamics (HZD) improved how robots' stability and performance were analyzed. It used virtual constraints to restrict all the joints and provided stable and coordinated motions for the robot. This approach simplifies control systems for the robots, ensuring they moved more reliably (Westervelt et al., 2018). As computers became more powerful, optimization-based controls became popular. This involves using computer algorithms to calculate the best possible movement paths for the robot while considering its physical limitations. However, these calculations are complex and can take a long time to converge, so they are often done in computer simulations before the robot is even turned on. To help robots adapt in real-time, scientists developed techniques like model predictive control (MPC). This approach simplifies the robot to a single rigid body in the calculations, making it easier to quickly decide how the robot should move next. An example of this in action is the Mini-Cheetah robot, which can walk, trot, and even gallop (Kim et al., 2019).

Despite these advancements, there's still a challenge in getting robots to adapt to new tasks or environments without human help. Machine learning algorithms have been used to teach robots different gaits, just like animals have different ways of moving at various speeds. For example, the ANYmal robot (Hutter et al., 2016) and the Raibo robot have shown impressive movements over tough terrains, as they have been trained using simulation-based techniques (Choi et al., 2023). However, these methods require substantial amounts of training data specific to the hardware in question and also demand considerable computational time and power to train the control policy effectively. They also face the challenge of transferring what is learned in a simulated environment to the real world, as there can be differences such as the moment of body inertia or mechanical frictions that are hard to predict. Additionally, if a robot encounters a situation it wasn't trained for, it may not know how to react, leading to potential damage or malfunction.

GAITS OF LEGGED ANIMALS AND ROBOTS

When observing a dog in motion, it's noticeable how its leg patterns vary with speed. This coordination of leg movements, known as a *gait*, is a common behavior among all legged animals, including humans. For example, humans typically walk with a phase where both feet touch the ground, transitioning to a run at higher speeds where both feet are momentarily airborne. Dogs, having four legs, display a wider range of gaits, including walking, trotting, and galloping.

Why do animals employ multiple gaits? It's widely believed that animals switch gaits to optimize energy efficiency over long distances (Hoyt & Taylor, 1981). Each gait is most energy-efficient at a specific speed range, akin to how a car shifts gears for fuel efficiency. Walking, for instance, maintains continuous ground contact, limiting stride length but reducing peak reaction forces from the ground. In contrast, trotting and galloping incorporate flight phases in which no part touches the ground, allowing for extended strides unrestricted by body and limb dimensions. Researchers Hoyt and Taylor (1981) quantified the energy expenditure of horses across different gaits and speeds, discovering that each gait minimizes energy use, or *cost of transport*, within certain speed ranges for horses. Beyond energy efficiency, recent studies suggest that gait changes enhance maneuverability and predator evasion, as observed in bipedal rodents like jerboas (Moore et al., 2017). Fully grasping the dynamics of these gaits is complex, considering the various footfall patterns and the intricate anatomy of animal skeletons. In the 1960s, Milton Hildebrand introduced a novel approach to study gaits, focusing on the timing and coordination of footfalls rather than just naming the gaits. He developed *gait diagrams* to classify animal movements into symmetrical and asymmetrical gaits, providing a visual framework for understanding these complex patterns (Hildebrand, 1965).

By emulating these natural gait patterns in the design of robot locomotion, we can design machines capable of adapting their movements to different speeds and terrains. Researchers have explored various methods to replicate these rhythms and transitions in robots. A particularly intriguing concept is that the synchronized limb movements in animals are governed by neural oscillations in their spinal cords, known as *central pattern generators* (CPGs). These neurons, interconnected and capable of exciting or inhibiting each other's oscillations, can settle into stable states that produce diverse rhythms for limb coordination (Duysens & Van de Crommert, 1998). This theory has been applied in robotics, as demonstrated by the salamander-inspired Pleurobot (Karakasiliotis et al., 2016). Moreover, the physical dynamics of a robot play a significant role in its movement capabilities. Even with predefined leg coordination and duty cycles (the proportion of each stride that a leg is in contact with the ground), robots like MIT's Mini-Cheetah can perform a variety of gaits, including walking, trotting, pacing, bounding, and galloping, across a wide speed range. This is achieved using MPC with specialized gait scheduling, showcasing the potential of robotic movement that closely mimics the natural world.

WHAT IS THE BEST FORM OF LOCOMOTION FOR SOCIAL ROBOTS?

The debate over whether robots should be equipped with legs and feet remains open. While legged locomotion may not always be the most efficient, particularly compared to speed of wheeled movement on flat surfaces or the unrestricted terrain access of flying, its advantages lie in a unique blend of versatility, adaptability, and efficiency. This is particularly evident in environments where wheeled and flying robots encounter limitations. Legged robots demonstrate unparalleled proficiency in navigating diverse terrains, from the rugged landscapes of distant planets to the chaotic and uneven grounds of disaster sites. Visualize a bicycle wheel rolling without its rim; each spoke can be equated to a leg. In this scenario, as the wheel turns, it resembles walking on countless legs, each following the other in rapid succession. The comparison of wheeled robots to legged robots (Asano & Suguro, 2012) highlights a key limitation: The inability of wheels to freely adjust in size or adaptively select footholds. In contrast, the limbs and legs of robots excel in these scenarios, adeptly overcoming obstacles and gaps impassable for their wheeled counterparts. The legged robot, in such scenarios, isn't dependent on continuous contact with the terrain for its movement. It possesses the agility to leap over gaps and can retract its legs to adeptly avoid colliding with obstacles. Furthermore, it has the capability to directly step onto raised platforms or steps. This versatility in navigating varied terrains and overcoming physical barriers sets it apart from other forms of robotic locomotion.

The physical design of legged robots, with their human-like ability to walk and climb, naturally fits into domestic and urban landscapes. Their movement through stairs and over obstacles mimics human actions, lending them an organic presence in spaces designed for people. This quality is invaluable in domestic settings, where robots need to move seamlessly within family homes, adapting to different room layouts and floor conditions. In emergency scenarios, like urban search and rescue, the agility of legged robots to traverse through debris and negotiate challenging terrains is a crucial asset. They can access areas that are otherwise hazardous or inaccessible to humans, making them indispensable in time-sensitive missions to locate and assist survivors. In healthcare environments, legged robots navigate crowded corridors and patient rooms with spatial constraints. Their ability to move at different heights and reach areas aids in various healthcare tasks, from delivering supplies to assisting in patient care, enhancing efficiency and support provided by medical staff. Furthermore, the interaction of legged robots with people and pets can be more natural, since their life-like movements allow them to integrate more harmoniously into social environments compared to the mechanical movements of other types of robots. This aspect is especially crucial in environments such as care homes, theatrical stages, or theme parks, where a robot's presence needs to be comforting and engaging, rather than intimidating or disruptive.

Ultimately, whether a robot should have legs and feet depends on its intended task and environment. There is a growing interest in transformable robotics in recent years, where a single robot can switch among different modes of movement; this potential further underscores the value of legged designs. An example is the ANYmal robot (Bjelonic, 2021), outfitted with detachable wheels, which adeptly shifts from walking to roller skating. This ability not only marries the stability and adaptability of legs with the speed of wheels but also opens new avenues for robotic applications. As the field of robotics rapidly advances, it is expected that a diverse array of robots will increasingly work alongside humans in various capacities. In industrial settings, these robots could collaborate with human workers, contributing to intricate tasks such as assembling components for cellphones or automobiles. Such robots possess the potential to seamlessly transition between different modes of movement to maximize efficiency in various tasks. They could use wheels to swiftly transport parts along assembly lines, ensuring speedy and efficient delivery. Once the delivery is complete, they can transform into legged mode to collaborate more effectively with human workers. This adaptability allows for a more dynamic and interactive work environment, where robots can adjust their form and function to suit the specific needs of the task at hand, be it rapid transport or precise, cooperative work in manufacturing processes. Their potential reaches well beyond the realm of manufacturing; these robots could also act as assistive companions, helping people with disabilities in ways similar to service animals. They could guide blind individuals on streets or within buildings, fetch items like newspapers, or serve as emotional or medical companions, offering support and assistance in daily activities. The integration of legged robots into daily life is set to become increasingly prominent. Their adeptness at navigating home environments renders them ideal for household assistance. This includes simplifying a range of tasks, from routine floor cleaning to more intricate chores such as washing dishes, making beds, or folding clothes, thereby enhancing convenience and efficiency in domestic settings. As these robots become more integrated into various aspects of daily life, their presence is set to evolve from a novelty to a necessity. The unique capabilities of legged robots, from their advanced mobility to their potential for human-robot collaboration with safety, position them as a crucial component in the ongoing advancement of robotics, reshaping how tasks are performed and services are provided in numerous fields.

REFERENCES

Asano, F., & Suguro, M. (2012). Limit cycle walking, running, and skipping of telescopic-legged rimless wheel. *Robotica, 30*(6), 989–1003.

Bjelonic, M. (2021). *Planning and control for hybrid locomotion of wheeled-legged robots* [Doctoral dissertation, ETH Zurich]. ETH Zurich Research Collection.

Choi, S., Ji, G., Park, J., Kim, H., Mun, J., Lee, J. H., & Hwangbo, J. (2023). Learning quadrupedal locomotion on deformable terrain. *Science Robotics, 8*(74), eade2256.

Duysens, J., & Van de Crommert, H. W. A. A. (1998). Neural control of locomotion; Part 1: The central pattern generator from cats to humans. *Gait & Posture, 7*(2), 131–141.

Gong, Y., Hartley, R., Da, X., Hereid, A., Harib, O., Huang, J. K., & Grizzle, J. (2019). Feedback control of a Cassie bipedal robot: Walking, standing, and riding a Segway. *Proceedings of the American Control Conference*, 4559–4566.

Hildebrand, M. (1965). Symmetrical gaits of horses. *Science, 150*(3697), 701–708.

Hoyt, D. F., & Taylor, C. R. (1981). Gait and the energetics of locomotion in horses. *Nature, 292*(5820), 239–240.

Hubicki, C., Abate, A., Clary, P., Rezazadeh, S., Jones, M., Peekema, A., Van Why, J., Domres, R., Wu, A., Martin, W., Geyer, H., & Hurst, J. (2018). Walking and running with passive compliance: Lessons from engineering: A live demonstration of the ATRIAS biped. *IEEE Robotics and Automation Magazine, 25*(3), 23–39.

Hutter, M., Hoepflinger, M., Remy, C. D., & Siegwart, R. (2012). Hybrid operational space control for compliant legged systems. *Proceedings of Robotics Science and Systems* (pp. 129–136). MIT Press.

Hutter, Marco, Gehring, C., Jud, D., Lauber, A., Bellicoso, C. D., Tsounis, V., Hwangbo, J., Fankhauser, P., Bloesch, M., Diethelm, R., & Bachmann, S. (2016). ANYmal—A Highly Mobile and Dynamic Quadrupedal Robot. *Proceedings of the International Conference on Intelligent Robots and Systems* (pp. 38–44). IEEE.

Kajita, S., & Espiau, B. (2008). Legged robots. In B. Siciliano & O. Khatib (Eds.), *Springer Handbook of Robotics* (pp. 361–389). Berlin, Heidelberg: Springer.

Karakasiliotis, K., Thandiackal, R., Melo, K., Horvat, T., Mahabadi, N. K., Tsitkov, S., Cabelguen, J. M., & Ijspeert, A. J. (2016). From cineradiography to biorobots: An approach for designing robots to emulate and study animal locomotion. *Journal of The Royal Society Interface, 13*(119), 20151089.

Kim, D., Di Carlo, J., Katz, B., Bledt, G., & Kim, S. (2019). Highly dynamic quadruped locomotion via whole-body impulse control and model predictive control [preprint]. <https://arxiv.org/abs/1909.06586>

Moore, T. Y., Cooper, K. L., Biewener, A. A., & Vasudevan, R. (2017). Unpredictability of escape trajectory explains predator evasion ability and microhabitat preference of desert rodents. *Nature Communications, 8*(1), 440.

Raibert, M. H. (1986). *Legged robots that balance.* MIT Press.

Sakagami, Y., Watanabe, R., Aoyama, C., Matsunaga, S., Higaki, N., & Fujimura, K. (2002). The intelligent ASIMO: System overview and integration. *Proceedings of the International Conference on Intelligent Robots and Systems* (pp. 2478–2483). IEEE.

Silva, M. F., & Machado, J. A. T. (2006). An overview of legged robots. *Proceedings of the International Symposium on Mathematical Methods in Engineering* (pp. 1–40).

Spagna, J. C., Goldman, D. I., Lin, P.-C., Koditschek, D. E., & Full, R. J. (2007). Distributed mechanical feedback in arthropods and robots simplifies control of rapid running on challenging terrain. *Bioinspiration & Biomimetics, 2*(1), 9–18.

Westervelt, E. R., Grizzle, J. W., Chevallereau, C., Choi, J. H., & Morris, B. (2018). *Feedback control of dynamic bipedal robot locomotion*. CRC Press.

CHAPTER FIVE

Wings & Propellers: Caring with/for Flying Robots

MAFALDA GAMBOA

On the slightly irregular wooden floor, two tripods were precisely positioned. We moved carefully around them—a small deviation and the tracking would not work as precisely. By the window, the two of us sat by a desk deliberately sorting pieces of technology. There were piles of small, numbered cardboard boxes neatly arranged on the desk. In stark contrast with the smoothness of the wooden desktop, a large number of electronic components, batteries, and propellers were laid out ready to be assembled. "I hope you do not think I am too protective of my drones," he said. But I did not think so—the way he unpacked each small drone, the way his hands moved, carefully putting together each piece, trying each interaction—all his mannerisms around the probes were inspiring.[1] These ritualistic gestures of preparation were reminiscent of familiar tasks of domestic care. I felt a deep responsibility and honor in being trusted and allowed into his robotic sensibility.

WHAT ARE SOCIAL DRONES ANYWAY?

In this chapter, I will take you on a short journey through the idea of drones as design-things, in the form of robots that are already making an impact in

---

1 Portions of this chapter were drawn from the author's licentiate thesis (see Gamboa, 2023).

the world. The words you will read here are positioned strongly in my own and other's design practice. Propelled—pun intended—by the research agenda in my project, I chose to focus on understanding the ways that the Human-Computer Interaction (HCI) discipline considers and applies design knowledge, approaches, and methods when studying social drones. In my work, you will not find a converged definition of social drones. Generally, within design, it may be advisable to not draw strict definitions, but rather to work with boundaries that can move. In the words of Redström (2017):

> … we do not settle for just one definition of design not because we do not understand the essence of design, but because it is much more powerful to work with difference as a basis when coping with complexity and change. And to work on the basis of (making a) difference, we need alternatives, and we need diversity. This is still a conversation between us about what design is, but it is one centered on its potentials for change, not its eventual convergence. (Redström, 2017, p. 141)

Working with drones requires the ability to deal with quick change and a generous capacity for a lack of convergence. Hence, Redström's avoidance in defining design also works when attempting to define social drones. Rather than focusing on definitions, I am concerned with understanding what the impact of design knowledge is when re-framing the role of social drones in diverse ways. I align with agential realism (Barad, 2007), an approach to knowledge where entanglement of the observer and the observed is fundamental to understanding the nature of reality, emphasizing the inseparability of the knower and the known in the process of knowledge formation. None of my work relies on a demarcated understanding of what exactly a social drone is. Rather, a drone can be both social and something else at the same time, both of them being true. The whole drone can be social, or the way it moves, or its propellers can be agents of social interaction—it is important to keep a certain open-endedness—all depending upon the unique assemblage of each everyday situation.

As drones become commercially available and applied in a variety of social contexts, the perceptions and ideas surrounding their use will be very different—as well as what (social) drones will be defined and re-defined as. To this end, we need to consider many voices and how they relate to the technology. It is precisely through multiple accounts—often from the fringes rather than the central tendency—that the most inspiring design possibilities are formulated. "Given that designs can be appreciated from a number of different perspectives, and that different people may find different ways to engage and make meaning with them—or fail to do so—multiple, inconsistent and even incompatible accounts may all be equally true" (Gaver, 2014, p. 17).

Drones have an established research domain of their own in Human-Drone Interaction (HDI) studies. They are undeniably becoming integrated in society as tools in a variety of work practices, such as mining, energy engineering, forestry, cinematography, and police work (Ljungblad et al., 2021) and in leisure activities such as photography (e.g., Hildebrand, 2017). This ongoing and advanced development suggests that we need to understand the different types of drone used in society to guide future design. For example, new application areas are emerging, including non-military surveillance, navigation, delivery, and first-person view (FPV) flight for entertainment. There is also an increasing research interest in the sub-set area of social drones so-called domestic drones (Baytaş et al., 2019; Karjalainen et al., 2017; Obaid et al., 2020)—drones being used in inhabited environments such as small-scale public spaces, the home, and the workplace, leading to complex interactions in intimate settings.

However, the way drones are studied in the context of HDI is not yet entirely tailored to the complexity of the sociotechnical assemblages (i.e., the ever-changing entanglement of human and technological actors) created in the real-life use and design of drones. In research, there is a tendency to study drones as tools—but a utilitarian perspective does not cover the full spectrum of drone encounters, including how vulnerable people and other agents may feel when interacting with them. Design methods that are not seeking to optimize performance are particularly helpful here. For example, Herdel and colleagues (2022) present drones as "helpful," "amicable," "functional," "knowing," "sensational," "reliable," and "unusual." Suspiciously, there is a gap in recognizing the "creepy," the "unreliable," the "invasive," the "unsustainable," and the "unwanted" drone—all of these are current characteristics of drones as perceived by many in society. A society which consists of many groups of people who would usually be called bystanders. I will return to the idea of bystander by the end of this chapter, but for now, bear with me. After more than three years of working within HDI, in casual everyday conversation about my research, the narratives I hear are often of highly uncomfortable or invasive encounters people have had with drones. This is where the idea of *care* becomes relevant.

DIMENSIONS OF CARE

After years of researching the design of drones, I have come in contact with many stories and examples of how they are used and cared for/with. Puig de la Bellacasa (2015) defines care as an ethico-onto-epistemological practice that involves attending to and valuing relations, entanglements, and the well-being of both human and non-human entities. I will use some stories to give examples of those relational worlds. I have throughout those years joined a number of online fora (for example, Facebook and Reddit) and been particularly attentive to news about drones. In the combination of the stories I have heard, the reactions I have seen online, and my previous work in autoethnography (Gamboa, 2022), I have found that there are many dimensions of care transmitted in the interactions with drones. Along with my children, I cohabited with a drone (labeled as appropriate for children) for over a year. Unexpectedly, the drone was more interesting as a fiddling device when idle than it ever was while flying. Our youngest daughter (who was 3 years old at the time) ended up leaving the room when the drone was in use, and ultimately when our third baby was born, I dropped the study in fear of harmful collisions. Drones challenge the notion of the user-bystander dichotomy, producing many layers of concern (from safety to legal) but more importantly they create worlds where neglect becomes evident.

I am inspired by what Latour (2004, 2005) describes as notion of the negotiation between matters of fact and matters of concern: Matters of fact are objective, verifiable information, while matters of concern encompass the broader, socially constructed issues that require collective attention and deliberation. More importantly in the context of this chapter, Puig de la Bellacasa (2015) steps further into the discussion of what she names "matters of care" (Puig de la Bellacasa, 2011) highlighting the ethical and relational dimensions of human and non-human interactions, emphasizing the interdependence and responsibility embedded in practices of care. Informed by Haraway's (2016) call to stay with the trouble, Puig de la Bellacasa (2011) "raises the issue of how 'we' are contributing to the construction of the world. How does respect for concerns in the things we re-present encourage attention to the effects of our accounts on the composition of things?" (p. 89). The age-appropriateness labeling of a toy drone is solely focused on a 1–1 relationship between one drone and one child—but what about their siblings? It is an act of care to take into consideration the "composition of things" in the stories we tell and the ways we design drones, to not simplify, and to stay with the trouble surfaced in these compositions and engage with them. The lack of consideration in the design for a match between the recommended age and appropriateness of use was clearly noted by the children: They could not independently change the batteries nor even press the "on" button as it was too fiddly. Was there caring consideration and engagement when designing this small button? The neglect for children as agents is noted in the lack of human-drone interaction research studies involving them.

Care is clearly not absent in existing human-drone relations. Drones often end up in accidents, either submerged or colliding with trees or even animals. Market leader DJI (n.d.) even offers an after-sales service curiously called "DJI Care Refresh: covering protection for collision, water damage, flyaway, and natural wear." This service offers, for a 1-year plan, "Two replacements in one year: Usable twice for damage, Usable once for flyaway" (DJI, n.d.). This is a quite significant service offering brand new drones at the cost of a replacement fee. In our previous work (Gamboa et al., 2023), we had asked in an online forum for short stories of drone accidents. In these stories we could identify many matters of care, from worries for privacy, drones returned after many years thanks to the kindness and curiosity of the finders, and collisions with other than human agents. In one instance:

> It flew away. I used the app litchi and had created a route it would follow, but it never came back. It turned out that I did not think that the stated height is in relation to the starting point and when the ground rose, it came closer and closer to the ground and finally stopped in front of a large spruce and did not know where it would go. There it hovered until the batteries ran out. A resident in the area found it after 1 1/2 years and when I checked the film on the memory card it was clear what had happened. (Anonymous Participant; in Gamboa et al., 2023, p. 5)

And in another:

> I'm on my 5th drone. I have crashed three drones and that is of course due to the way I use the drone. My drone is a camera dolly and the best movie clip is when you drive backwards and sideways. Unfortunately, you do not see in the direction of travel either. I usually film my grandson, who engages in kite surfing and kite foiling, 2–3 m above the water is usually the best and safest, but sometimes you end up below 1m and then there is a crisis. (Anonymous Participant; in Gamboa et al., 2023, p. 8)

Through these stories, I find it clear the need to de-center the human as the default agent in HDI. For example, birds, trees, or landscapes should not be seen as passive subjects to a user-bystander relationship between humans and drones. The development of this technology must include considerations towards the agencies of the world beyond the human ... and consider the marks those agencies leave. Once, while scrolling a forum for DJI drone owners in Sweden, there was a post that caught my attention. There was an image in it: A submerged drone in an old lime quarry. The post was written by a group of divers, and it read: "During the morning today we found a Mavic 2 drone. The owner had filmed the quarry, but even themselves. Help us find the owner so they can find their drone back. Share share share."

The initial post gathered more than 120 shares. There were a number of comments, many with suggestions on how to track the owner through the SD card inside the drone. Some asked if there was no operator ID, as this is usually required in Sweden. Many pointed out the propellers were gone, finding it suspicious. Some theorized that the drone was just dumped without them, while others thought perhaps a fish or crayfish had been the culprit. One week later, the diver's group posted a photo of the owner reunited with the drone. I can't help but be curious—what was the story? Who took the propellers? What happens to the drone now? Stories such as this show the complexity of the sociotechnical assemblage surrounding that one drone. Because these robots are in the wild interacting with the world, they have become invaluable objects to research. This story surfaces many dimensions of human care, which cannot be neglected, such as care for the drone, for the environment, and for the owner who may have been disappointed.

In our work, we are studying the possible application of rather large (a little over 2-meter wingspan) vertical take-off and landing (VTOL) drone to deliver blood tests and medicine between healthcare centers and laboratories (see photo 3 in Fig. 05.1). During one of those studies, we were invited to observe as the drone operator prepared for a test flight. As he went through his checklist, he moved around the drone, carefully prodding parts with his skillful fingers, checking for balance and blockages or malfunctioning parts. Eventually, as he prepared to connect the battery, he put forward a set of 3-D printed propeller guards, screwed onto a wooden bar—clearly a homemade solution. This guard was meant to make sure the propellers would not harm the operator when the drone turned

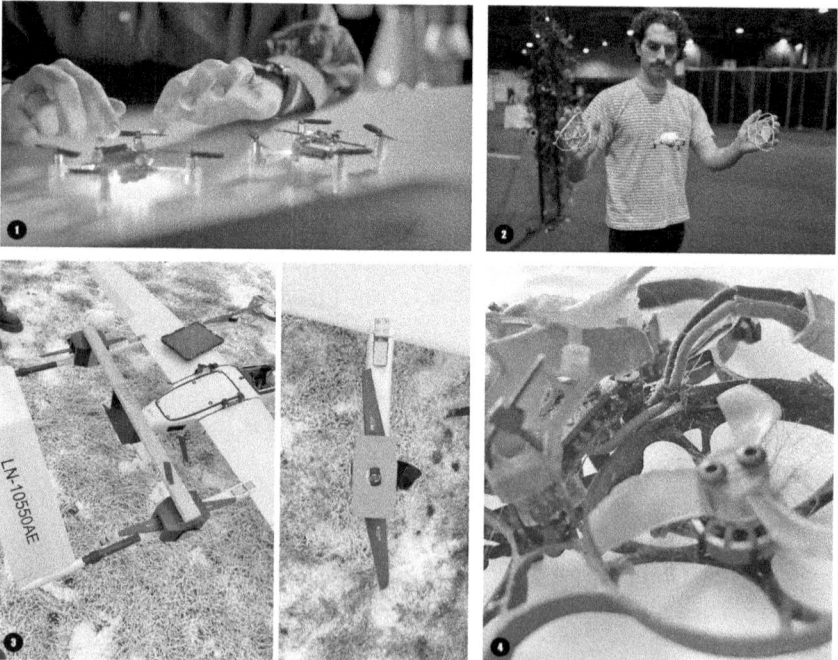

Figure 05.1: A set of images with a focus on drones and their propellers. (1) A close-up of me touching the propellers of the small drone I work with. Before each flight, I attentively check if each propeller is rotating properly and attached to the motors. (2) Joseph la Delfa controlling his drone. (3) A fixed wing large drone and its home-made propeller guard. (4) A self-built hobby drone with zip-ties instead of screws to facilitate the replacement of broken parts. Note how there is hair tangled on the propellers (source: author and Joseph La Delfa).

on. The roughness of this safeguard came in strong contrast with the sleekness of the drone—a cheap plastic orange 3-D printed shape encasing the expensive and sharp-looking propellers. I could not help but think of the careless damage those could cause up in the air—damage to anyone or anything other than the pilot.

In a series of interviews I am conducting with hobby drone pilots, I have encountered an avid indoor first-person drone builder and driver. He makes his own obstacle tracks in his home and proudly exhibits his machines in his living room. As he told me the story of his flights, he let me hold the drone, he told me of the cuts in his fingers, of the time he had to wait for his girlfriend to help him disentangle the drone from his hair (see photo 4 in Fig. 05.1). It suddenly became clear to me the blurred boundaries between him and the drone: The machine left marks in him, and he left marks in it. There is no way of fully understanding these interactions without considering the objectivity that lies in the dissolving of the edges of the body. Note how in the example above, the hair becomes part of the drone, as much as the drone became part of his hair. In the words of Karen Barad (2007, p. 824), "Either way, what is important about causal intra-actions is the fact that marks are left on bodies. Objectivity means being accountable to marks on bodies."

PROPELLING CARE

It is in the context of accountability that the encouragement of multiple interpretations of drones should take place. It is—as I see it—our duty and role to put forward the fringe views at risk of being hidden. As encounters with drones become more commonplace, it is increasingly important for us to turn our attention to the everyday interactions we have with them. Are we truly, as a society, willing to give the space above our heads and around our bodies to be populated by flying robots? Here, I will finally return to the notion of bystander in human-drone interaction. I have throughout the years struggled with assigning the label of bystander to anyone outside the drone-pilot duo. It took me years to properly articulate my reticence, but I see it now as a caring gesture. It is a matter of care to prevent people from assuming they have no agency in their interactions by avoiding the use of the word bystander. A bystander can be a citizen, a lawmaker, a politician, a school child, an annoyed girlfriend, a mother, a terrified person, a nurse, a dog, a flowing river. Placing all of these agents under the umbrella of bystander—a word that ultimately means witnessing something but not being part of it—is neglecting the importance of being part of the network of care. Revisiting Redström's resistance to definitions: The definition of bystander is less helpful than it would be to carefully identify the involved parts and what agencies are possible in the sociotechnical assemblage.

Now traveling back to the prologue in this chapter, I tell the story of a short interaction with Joseph la Delfa. He has done work in developing unlikely drones, such as Drone Chi (La Delfa et al., 2020)—in which he studied the practice of tai chi to inspire a design for moving with drones. In Fig. 05.1 (photo 2) you can see him holding his hands around his drone. The careful assembly of these fragile machines was a delicate procedure where pieces would only fit if placed exactly in order. Working with those small drones consistently damaged me in small ways—replacing broken propellers broke my nails—both mine and the drone's body together becoming and making marks in one another. For us, it was our nails, but for many others the implications are larger. In all scales, drones will change the bodies of all actors they interact with, in more or less harmful ways. Understanding how the assemblages can be beneficial or detrimental is a necessary work of care. Drones will, through their widespread use, exaggerate already-existing schisms and inequalities. If we fully consider the responsibility of making visible the impact of these machines on bodies, we can exercise a necessary work of propelling acts of care toward the world. As we care for them, as we pick them out of the bottom of the lake, as we interweave our hair and our privacy with their cameras, propellers, and wings, we give in to a future we may not want—not amicable or functional, but rather unsustainable and invasive.

REFERENCES

Barad, K. (2007). *Meeting the universe halfway: Quantum physics and the entanglement of matter and meaning.* Duke University Press.

Baytaş, M. A., Çay, D., Zhang, Y., Obaid, M., Yantaç, A. E., & Fjeld, M. (2019). The design of social drones: A review of studies on autonomous flyers in inhabited environments. In *Proceedings of the Conference on Human Factors in Computing Systems* (pp. 1–13). ACM.

DJI. (n.d.). DJI Care Refresh. <https://www.dji.com/se/support/service/djicare-refresh>

Gamboa, M. (2022). Conversations with myself: Sketching workshop experiences in design epistemology. In *Proceedings of Creativity and Cognition* (pp. 71–82). ACM.

Gamboa, M. (2023). *From the ground up: Designerly knowledge in human-drone interaction* [Licentiate thesis, Chalmers University of Technology]. <https://research.chalmers.se/en/publication/534696>

Gamboa, M., Ljungblad, S., & Sturdee, M. (2023). Conversational composites: A method for illustration layering. In *Proceedings of the International Conference on Tangible, Embedded, and Embodied Interaction* (pp. 1–13). ACM.

Gaver, W. (2014). Science and design: The implications of different forms of accountability. In J. S. Olson & W. A. Kellogg (Eds.), *Ways of knowing in HCI* (pp. 143–165). Springer New York.

Haraway, D. J. (2016). *Staying with the trouble.* Duke University Press.

Herdel, V., Yamin, L. J., & Cauchard, J. R. (2022). Above and beyond: A scoping review of domains and applications for human-drone Interaction. In *Proceedings of the CHI Conference on Human Factors in Computing Systems* (pp. 1–22). ACM.

Hildebrand, J. M. (2017). Situating hobby drone practices. *Digital Culture & Society, 3*(2), 207–218.
Karjalainen, K. D., Romell, A. E. S., Ratsamee, P., Yantac, A. E., Fjeld, M., & Obaid, M. (2017). Social drone companion for the home environment. In *Proceedings of the International Conference on Human Agent Interaction* (pp. 89–96). ACM.
La Delfa, J., Baytaş, M. A., Patibanda, R., Ngari, H., Khot, R. A., & Mueller, F. F. (2020). Drone Chi: Somaesthetic human-drone interaction. In *Proceedings of the CHI Conference on Human Factors in Computing Systems* (pp. 1–13). ACM.
La Delfa, J., Baytaş, M. A., Luke, E., Koder, B., & Mueller, F. F. (2020). Designing Drone Chi. *Proceedings of the ACM Designing Interactive Systems Conference* (pp. 575–586). ACM.
Latour, B. (2004). Why has critique run out of steam? From matters of fact to matters of concern. *Critical Inquiry, 30*(2), 225–248.
Latour, B. (2005). *Reassembling the social: An introduction to Actor-Network Theory*. Oxford University Press.
Ljungblad, S., Man, Y., Baytaş, M. A., Gamboa, M., Obaid, M., & Fjeld, M. (2021). What matters in professional drone pilots' practice? An interview study to understand the complexity of their work and inform human-drone interaction research. *Proceedings of the CHI Conference on Human Factors in Computing Systems* (no. 159). ACM.
Obaid, M., Johal, W., & Mubin, O. (2020). Domestic drones: Context of use in research literature. *Proceedings of the International Conference on Human-Agent Interaction* (pp. 196–203). ACM.
Puig de la Bellacasa, M. (2011). Matters of care in technoscience: Assembling neglected things. *Social Studies of Science, 41*(1), 85–106.
Puig de la Bellacasa, M. (2015). Making time for soil: Technoscientific futurity and the pace of care. *Social Studies of Science, 45*(5), 691–716.
Redström, J. (2017). *Making design theory*. MIT Press.

CHAPTER SIX

Color & Clothing: The Social Consequences of Aesthetics

NATALIE FRIEDMAN

Aesthetic choices made in the design of robots play a significant role in how they are perceived at first impression and interact with humans. While functionality and efficiency of robots remain paramount, aesthetics—particularly color and clothing—have gained recognition in the human-robot interaction (HRI) community as essential aspects of their design. Clothing for robots has been appearing in science fiction for a long time: Rosie from *the Jetsons* wearing an Apron, Baymax from *Big Hero Six* wearing armor when getting ready for a fight. We also see coverings for machines (e.g., phone cases, decals for Roomba) which aren't so far away from these science fiction examples. In this chapter, we will explore the profound implications of color and clothing for robots, focusing on aesthetics and their social consequences. Aesthetics matter in robotic clothing but, perhaps counter-intuitively, anthropomorphism is not necessarily the goal.

Before delving into the aesthetics and social implications, I will establish definitions of clothing and color in the context of robots, so we have a foundation of understanding. Clothing refers to any external covering or material layer that a robot wears or that is utilized as part of its design. If designers want robots to fit into a variety of environments, clothes should help a robot be adaptable, protected, and socially signal so that the robot can fit in appropriately to a given context. Clothes should be designed for what the robot needs, not what our idea of clothing should do in general (Friedman et al., 2021). For example, a four-legged robot might not make much use of pants that cover only two legs. Clothes can include removable protective shells, outer casings, or any form of fabric or material outside of the robot's construction. Color of a robot is determined by the material it is made with. Common materials that robots are constructed with are plastic, wood, and metal (McComb, 2003). Each of these materials can be painted but show a different shade and texture depending on the original material. Clothing can also have color that socially signals a role of a robot. The choice of color should be deliberate, since color influences human emotions (Terwogt & Hoeksma, 1995) and can contribute to interpretations of machine agents' purpose and character. For instance, color can influence perceptions of robot task or intention (Pörtner et al., 2018) and can even parallel human racial cues that carry problematic or benign stereotypes (Sparrow, 2019).

SETTING FIRST IMPRESSIONS: THE ROLE OF COLOR IN AESTHETICS

Much like in human interactions, the color of a robot's body and clothing is part of the first impression it makes on observers. This initial perception can shape human-robot interactions, influence trust (Xu & Howard, 2018), and convey a robot's intended role (i.e., a nurse might wear a light blue whereas a janitor might wear a dark blue) or personality (i.e., an extroverted robot might wear bolder colors). In an increasingly diverse world of robots, color can provide critical cues about a robot's function. A robot's clothing color can indicate whether it is for industrial tasks (i.e., bright orange for visibility) or household chores (i.e., aprons with colorful patterned prints to blend in). We pick up on these cues because of our experiences in the world with people, as we more generally draw on human norms to make sense of robot cues and behaviors (Edwards et al., 2016). An industrial robot's body or clothes may feature a robust and dark-colored exterior, symbolizing strength and durability, while a medical assistance robot might incorporate soft-pastel scrubs, signifying gentleness and care as well as skill and knowledge, as a nurse would similarly wear in the USA (Thomas et al., 2010). In some cases, being conspicuous is essential for robotic safety and effectiveness. High-visibility colors, such as fluorescent orange or neon green, can make a robot stand out in its environment, reducing the risk of accidents and ensuring that humans are aware of its presence (see Zielinska et al., 2017). This is particularly important in settings where robots and humans collaborate closely, such as construction sites or warehouses.

The perception of color in clothing is deeply rooted in historical contexts. Robots are often perceived as social actors (cf. Reeves & Nass, 1996) and therefore should be dressed with social awareness in mind. For robots to fit into society, designers should understand these historical nuances when designing for diverse audiences. Throughout history, colors have carried different meanings. For instance, as early as 1500 B.C. the color purple once symbolized royalty and wealth, signifying power and prestige because of the expense to make purple dye (Ashby, 2007). In fact, the production of "Tyrian purple" once involved tens of thousands of snails and a lot of labor (Cartwright, 2016). In modern times, it is no longer expensive to make this dye and therefore does not symbolize wealth. In some cultures, specific colors may signify different values. For example, white is associated with purity and innocence in Christianity (Ureña et al., 2022), while in the medical field, it signifies sterility and cleanliness. Designers of multi-use robots must consider the cultural significance of color to ensure appropriate and respectful interactions. Therefore, when designing clothing for robots for a global audience, understanding these cultural nuances becomes paramount to ensure effective communication and cultural sensitivity.

ADAPTABILITY THROUGH CHANGING CLOTHING

Designers can change a robot's clothing, altering its color and appearance to suit different contexts or tasks. This adaptability—not unlike humans' ability to change self-presentation for different contexts and goals (Goffman, 1959)—allows robots to seamlessly transition between roles and environments. Elsewhere (Friedman et al., 2021), my colleagues and I have explored the multifaceted roles of clothing for robots. I've highlighted two noteworthy aspects: adaptability through social signaling and protection.

Colors in robotic clothing can have pragmatic applications beyond aesthetics and cultural considerations. As noted, they can serve as a means of practical identification and differentiation. Because two identical robots may have different roles, like waiter or a bus boy in a restaurant, bystanders need to be able to differentiate them. In fact, at SoftBank, the company that created Pepper, the head of marketing explained, "At first, I thought it was crazy to have robots wearing clothes! But then I realized how important it is to distinguish between Peppers with identical faces" (Nippon.com, 2017). For example, in a hospital setting, different colored uniforms for various medical specialties can help patients and staff easily identify the roles and expertise of robotic healthcare providers.

A notable case study in the pragmatic use of color in robotic clothing involves the application of thermochromic fabric. Thermochromic ink changes color at a specific temperature and is often applied to textiles and smart windows (Thamrin et al., 2022). For robots, this ink can provide real-time feedback on a robot's internal state, particularly its temperature. Imagine a scenario where a robot is assisting in a high-temperature environment, such as a kitchen. As the robot performs tasks, its thermochromic fabric gradually shifts from a cool blue to a fiery red, indicating that the robot's body becoming too hot. This color change acts as an intuitive signal for both the robot's operators and nearby humans, prompting them to take necessary actions such as providing cooling or adjusting the robot's workload.

RECONSIDERING HUMAN NOTIONS OF CLOTHING

While we have the opportunity of the adaptability affordance in clothing for robots, we need to think critically about the relationship between colored clothing and the context of the robot. There are four potential divergences that deserve consideration: Anthropomorphism, roles, and design.

Clothing Need Not Necessarily Be Anthropomorphizing

Clothing can allude to anthropomorphic forms and materials, but does not need to be explicitly pants, glasses, or a shirt. Sometimes, dressing robots like people can take away from the overall human-robot interaction. More specifically, making a robot look like a person could increase expectations of robots, making a bystander think it a robot could function as a human might. That is, for a human-like robot, "People expect them to adhere to human norms and have much higher expectations regarding their capabilities compared to robots with machine-like appearance" (Złotowski et al., 2015, p. 356).

Also, if a robot's main function is something like vacuuming or assembling cars, having a human-like form might not be necessary, as a robot vacuum does not need arms. So, it really depends on the task at hand. For social tasks, anthropomorphic-leaning clothes might help people understand what the robot is for and therefore help a robot do its job. Ultimately, the designer must consider what the needs of the robot are, whether they be social or functional, and design for those needs.

Scholar and designer Ylva Ferneaus studies ways to communicate robot emotions and stories through prototyping removable labels on a robot (Fernaeus & Jacobsson, 2009). While labels are not explicitly clothes, they do have features of aesthetic cueing and removability, similar to clothes. For example, Fernaeus writes the words, "I'm shy" on a Roomba when it is under the couch. While the idea of shyness is anthropomorphic, she indicates shyness through words instead of facial expression. Potentially, by using words instead of forms, the robot might be less likely to generate discomfort among humans via the uncanny valley problem (Mori et al., 2012), which is when a robot looks almost human but not quite, it can leave people feeling uneasy or as though the robot is creepy. Using clothing-like words, forms, and materials that allude to robot states in ways that humans can intuitively understand, robots can be augmented to communicate what they might do next without leading to explicit expectations that they can see or hear like humans do.

Clothing Can Exhibit Robot Updates

For people, clothing demonstrates what has been and what is to come: A past experience (e.g., a wristband from a concert), an emerging identity (e.g., a collared shirt for a job interview), or a signifier of a mood (e.g., dark clothing for a sad or angry state). Fashion as a way of temporarily expressing what the wearer is becoming could apply to the future of clothes for robots. Robots often have internal software updates, which are often not displayed on the external hardware. Clothes could demonstrate what type of software they will have (Friedman, 2023). For example, a new version of software that allows robots to move faster might be indicated by clothes that are a deeper red or a pin with a speeding symbol. From this, a bystander might know not to get as close just yet, until they are aware of the new speed.

CHALLENGING TRADITIONAL GENDERING THROUGH ROBOT ATTIRE

People tend to gender robots and behave differently around robots based on their inferred gender. Siegal and colleagues (2009) found that men tended to be more persuaded by female robots and both genders tended to find the robot of the opposite sex more credible. This tells us that a gendered appearance has a strong impact on human-robot interaction (see Chapter Eighteen). Thus, we must take a closer look on how gendered clothing impacts human-robot interaction.

The introduction of clothing for robots opens up a new frontier where attire might not be confined by gender norms as clothing tends to be for humans. Traditional human clothing often carries gender-specific connotations, but for robots, I believe there is an opportunity to create a non-gendered category of clothing both because of its form and because of its ontological status. While some may argue that a round robot might appear more feminine and a jagged edged robot more masculine, robot forms can be non-gendered. Non-gendered robot forms could challenge the stereotypical forms and colors associated with gender-specific attire. Robots designed for various roles, irrespective of whether they're intended for caregiving, maintenance, or social interaction, could sport attire that doesn't adhere to conventional gender distinctions. Designers can also create robots engaging traditionally female roles to wear male clothing forms in order to challenge previous biases humans have, expanding our perception of what gender can be. For example, Baymax from *Big Hero Six* is a healthcare robot, and while nurses are traditionally women, Baymax has a soft male voice and an ambiguous form. However, in a fighting scene, Baymax wears armor which has a male silhouette. Baymax responds, "I'm concerned this will undermine my non-threatening huggable design" (Hall & Wiliams, 2014). Baymax doesn't necessarily immediately embrace this new, traditionally masculine form. This scene challenges gender norms, as a robot who is supposed to be male, prefers to maintain his caretaker role, which is traditionally a female role. In the bigger picture, these presentations could promote inclusivity and diversity in robot design, fostering a more neutral and egalitarian approach to their appearance.

Designing Clothing and Robots in Tandem

In my interviews with people who dress robots, experts explained the frustration that clothes were designed *after* the robots were designed and built. Stacey Dryer, a toy designer and clothing design for Dragonbot, a social robot made for kids, explained, "Because the mechanics of the robot were pretty much complete by the time I came on board, there were many challenges such as keeping the fabric away from any pinch points that needed to be addressed solely by the skin's construction. Whereas in the toy industry mass market many of these elements like key attachment points and buffer zones are worked into the inner structure." She points to the toy industry as a standard because it considers forms and clothing in relationship to those forms early on.

Through this lens, designing clothes for robots should be an integral part of the overall design process rather than an afterthought. Designing robots and their clothing *in tandem* ensures that the body and clothing complement each other. By considering the robot's social requirements, context, functionalities, movement capabilities, and specific needs from the outset, clothing can be tailored to enhance not only the robot's aesthetic appeal but also its functionality and interaction with humans. Such coordinated design can prevent issues such as hindering a robot's movement or impeding its sensors due to ill-fitted or poorly designed clothing. In addition to designing early, clothes should be updated consistently as the robot enters new contexts in order to help a robot fit in socially and physically. This approach can prevent a robot being too large for its environment or being inappropriately dressed, ultimately helping a robot stay active in new contexts.

PRAGMATICS: COULD ROBOTS DRESS THEMSELVES?

While I would recommend considering the social context and roles of a robot before dressing it, I also would recommend testing the intended clothes to learn about the pragmatics of dressing. To demonstrate, I experimented with a Kinova Gen3 arm to learn how a robot could dress itself. Before it could dress itself autonomously, I wanted to see if I (someone who knows how clothing works) could control its inherent mechanics to functionally handle the act of dressing. So I first teleoperated this robot arm, with its seven degrees of freedom, attempt to put clothing on it using only its own parts and functions. It was nothing short of awkward. I repurposed a human-pant leg as a sleeve and tried to teleoperate the robot to take it off. The robot couldn't pull the clothes off with its gripper. I tried to use gravity to get the pants off, shifting the robot arm upside down so the pants would fall off. The robot couldn't be fully upside down and the clothes would get bunched up around the gripper. Finally, one strategy worked. I teleoperated the gripper to close around a ribbon and made the robot arm rotate on its horizontal axis while the ribbon wrapped (or unwrapped, depending on the rotation direction) around the arm. This was successful, but still very slow. Teleoperation, limited degrees of freedom, and thick material contributed to my challenges. In the future, I can see a robot that is designed to move in a way where it has enough awareness of its own form to dress itself. Smooth and stretchy material also could be a standard for dressing robots, so the fabric can glide on and off the robot and pull without damaging the robot.

While the pragmatics are a hurdle to get over, I will abandon that reality right now and present to you a design fiction. Imagine a future in which robots have the agency to dress themselves. This landscape introduces a combination of artificial intelligence, personalized preferences, and environmental awareness, allowing robots to curate their own attire. For example, perhaps they have a GPS and a social calendar which helps them know which occasion to dress for. Maybe they have a thermometer and access to forecasts that help them dress for the weather, so they don't overheat or get too cold. Perhaps each robot is designed to have a personality (which can support meaningful HRI; Robert et al., 2020), and it selects from a closet that is designed for that robot's style based on this personality. Imagine generative AI helps robots develop this style and a knitting machine works with a sewing machine to make the clothes, and a robot dresses another robot. This future could have major benefits. For one, a robot could demonstrate an appealing character novel to people, encouraging curiosity and interaction. Next, efficiency and safety could be improved; there would be no more need for a robot's dresser, as they would not have to don and doff the clothes, which can often be awkward and even dangerous (i.e., for ERICA, roboticists needed to remove the arms to put clothes on). Perhaps a robot that dresses itself could also be more precise—a bowtie might be perfectly straight and a shirt might be ironed without a wrinkle.

There may be initial costs, as people would need to redesign robots to move in new ways for a robot to dress itself. Every robot (and robot context) has different needs from clothes, which means there may need to be different customization for each robot. To mediate this, perhaps there could be a standard in which robots have software and hardware so they can dress themselves easily. As for software, this standard could be in the form of a plugin or could be customized for each robot. The robot could be designed to have more awareness of its own body. For hardware, a robot should have attachment points for clothes, and it should be seamless to don and doff. In other words, one should not have to take off an arm and a leg to put clothes on the body. Static hardware features could also reduce the need for customization. For example, a screen or LEDs, which show dynamic displays can change styles or colors based on the context. Other clothes that could have static forms and dynamic colors are cloth with thermochromic ink, while shape memory alloy could change form based on the temperature.

A potential social risk of a robot dressing itself could be missing a context clue if the sensors (GPS, thermometer, calendar, etc.) don't pick up a social cue. To address this, a human who is more sensitive to the culture and context could select the clothes, improving the possibility that the robot and its attire are fit to the social context. Another risk of a robot dressing itself is maintenance and repair. Clothing gets dirty, it can rip, or get stretched out. In future work we can consider how robots would handle maintenance or repair of their clothing, potentially integrating self-healing materials or mechanisms for self-maintenance. While the emergence of self-dressing robots could provide many opportunities for novel style, people should still oversee the robot's choices to account for social contexts. Considering cultural sensitivity, and maintenance frameworks could pave the way for a more comprehensive and socially adaptable future.

REFERENCES

Ashby, N. (2007). *Simply color therapy*. Zambezi Publishing.

Cartwright, M. (2016). *Tyrian purple*. World History Encyclopedia.

Fernaeus, Y., & Jacobsson, M. (2009). Comics, robots, fashion and programming: Outlining the concept of actdresses. In *Proceedings of the International Conference on Tangible and Embedded Interaction* (pp. 3–8). ACM.

Friedman, N. (2023). *Designing clothing to improve human-robot interaction*. [Unpublished doctoral dissertation]. Cornell University, Ithaca, NY, USA.

Friedman, N., Love, K., LC, R., Sabin, J. E., Hoffman, G., & Ju, W. (2021). What robots need from clothing. In *Proceedings of the Designing Interactive Systems Conference* (pp. 1345–1355). ACM.

Goffman, E. (1959). *The presentation of self in everyday life*. Doubleday.

Hall, D., & Williams, C. (Directors). (2014). *Big hero 6* [Film]. Walt Disney Pictures.
Edwards, C., Edwards, A., Spence, P. R., & Westerman, D. (2016). Initial interaction expectations with robots: Testing the human-to-human interaction script. *Communication Studies, 67*(2), 227–238.
McComb, G. (2003). *Robot builder's bonanza*. McGraw-Hill.
Mori, M., MacDorman, K. F., & Kageki, N. (2012). The uncanny valley [from the field]. *IEEE Robotics & Automation Magazine, 19*(2), 98–100.
Nippon.com. (2017, Feb. 27). Robot runway: Pepper's fashion show debut. *Nippon News*. <https://www.nippon.com/en/views/b00911/>
Pörtner, A., Schröder, L., Rasch, R., Sprute, D., Hoffmann, M., & König, M. (2018). The power of color: A study on the effective use of colored light in human-robot interaction. In *Proceedings of the International Conference on Intelligent Robots and Systems* (pp. 3395–3402). IEEE.
Reeves, B., & Nass, C. (1996). *The media equation: How people treat computers, television, and new media like real people*. Cambridge University Press.
Robert, Jr., L. P., Alahmad, R., Esterwood, C., Kim, S., You, S., Zhang, Q., et al. (2020). A review of personality in human–robot interactions. *Foundations and Trends in Information Systems, 4*(2), 107–212.
Siegel, M., Breazeal, C., & Norton, M. I. (2009). Persuasive robotics: The influence of robot gender on human behavior. In *Proceedings of the International Conference on Intelligent Robots and Systems* (pp. 2563–2568). IEEE.
Sparrow, R. (2019). Do robots have race?: Race, social construction, and HRI. *IEEE Robotics & Automation Magazine, 27*(3), 144–150.
Terwogt, M. M., & Hoeksma, J. B. (1995). Colors and emotions: Preferences and combinations. *The Journal of General Psychology, 122*(1), 5–17.
Thamrin, E., Warsiki, E., Bindar, Y., & Kartika, I. (2022). Thermochromic ink as a smart indicator on cold product packaging-review. *IOP Conference Series: Earth and Environmental Science, 1063*, 012021.
Thomas, C. M., Ehret, A., Ellis, B., Colon-Shoop, S., Linton, J., & Metz, S. (2010). Perception of nurse caring, skills, and knowledge based on appearance. *JONA: The Journal Of Nursing Administration, 40*(11), 489–497.
Ureña, L. G., Valeriani, E., Angelini, A., Carretero, C. S., & Gimeno, M. S. (2022). *The language of colour in the Bible: Embodied colour terms related to green* (Vol. 11). Walter de Gruyter GmbH & Co KG.
Xu, J., & Howard, A. (2018). The impact of first impressions on human-robot trust during problem-solving scenarios. In *Proceedings of the International Symposium on Robot and Human Interactive Communication* (pp. 435–441). IEEE.
Zielinska, O. A., Mayhorn, C. B., & Wogalter, M.S. (2017). Connoted hazard and perceived importance of fluorescent, neon, and standard safety colors. *Applied Ergonomics, 65*, 326–334.
Złotowski, J., Proudfoot, D., Yogeeswaran, K., & Bartneck, C. (2015). Anthropomorphism: Opportunities and challenges in human–robot interaction. *International Journal of Social Robotics, 7*, 347–360.

CHAPTER SEVEN

Tactility & Texture: Embrace at the Interface

JASON EDWARD ARCHER

In 2005, roboticists from MIT's Media Lab presented research from their Robotic Life Group based at an international conference (Stiehl & Breazeal, 2005; Stiehl et al., 2005). They argued that developers of robotics systems, which at that point had primarily been built for industrial settings, had largely focused on object manipulation. This meant little attention had been paid to how robotic systems might interact with human users. In particular, tactile sensors were only considered for grips that manipulated objects and developers failed to consider how robots might physically feel to humans. The researchers argued that touch plays a vital role in supporting companionship and had potential health benefits as well—claims backed up by empirical research into human-human touch (Fields, 2001) and human-animal touch (Ballarini, 2003). The researchers developed Huggable, a robot that looked like a stuffed teddy bear, to address the lacuna of research on the affective, relational, and health impacts of human-robot interaction. Huggable was also meant to fill a technical gap left by other popular social robots at the time, including Paro and AIBO. As the authors pointed out, the seal-like Paro and dog-liked AIBO had limited touch sensors located only in certain parts of their bodies while Huggable included sensors over its entire body and the ability to actively "touch back through nuzzling, hugging, and other communicative touch behaviors" (Stiehl et al., 2005, p. 408).

The name Huggable evokes a desire to explore the affective (psychophysical, social, and cultural foundations related to feelings and emotions) and relational dimensions of social robots as touch oriented. After all, what could be more social than a hug? What Huggable represented was an early attempt to understand how touch matters in human-robot interactions that take seriously the issues of tactility and texture. Attempts to develop technical systems of touch for social robots and research to determine the social implications of these systems constitutes a growing area of academic and commercial inquiry, but it still lags behind technical and social research in areas related to visual, verbal, and auditory communication. Despite advances related to the technical features and social implications of touch for social robots, most studies focus on affective, relational, and health dimensions first articulated in the Robotic Life Group's research. While questions about how touch matters in these areas matter and should be expanded, there are provocative issues to explore regarding the texture and tactility of social robots that go beyond personal and interpersonal effects—issues that intersect around care, labor, gender, and power, to name a few.

MACHINE TOUCH: TACTILITY AND TEXTURE

Tactile systems[1] and the texture of robots are discrete yet interrelated technical components of social robots that allow them to touch, respond to touch, and to connect with humans through touch. Tactility is, simply, the ability for a robot to be touched and respond to touch. Texture is the surface characteristics of a robot that impact its perceived tactility, and the way a robot moves impacts the ways its textures are felt. Many tactile systems developed for robots are inspired by psychophysical research on human touch and are meant to emulate those systems to an extent. From a psychophysical orientation, human touch is made up of multiple physical and perceptual systems including proprioception and kinesthesia (awareness of one's own body), vestibular (sensations informing balance), and tactile. The tactile system relates to sensations of pressure that can lead to perceived sensations like vibrations, elasticity, and density. The tactile system in humans also includes fibers related to sensing temperature (thermoception), pain (nociceptors), and stimulating positive affect (C-tactile afferents; Paterson, 2023a). Extrapolating from research on human touch is used to develop different types of tactile components for robots but it is also motivated by a desire to design social robots that can affect certain kinds of reactions from human partners. Research on C-tactile afferents, in particular, offers evidence there is an optimal pressure and speed of stroking or caressing that is sensed by specialized fibers which

---

1   The term tactile is used here instead of haptic or other touch-related terms to denote the typical language used in robotic engineering and HRI literature, but the usage is often inconsistent and there is an important debate to consider regarding the genealogical development of the term tactile and haptic, including the ways they carry histories and normative assumptions (see Parisi, 2018).

stimulate positive affective reactions in humans (Pawling et al., 2017). The evidence is tantalizing in that it suggests the possibility of unlocking emotional connections between humans and robots through specific ways of touching, but the links between C-tactile afferents, the promotion of positive emotional responses, and increases in a sense of wellbeing are complicated and still not well understood (Schirmer et al., 2023). Untangling the connections will require accounting for intersecting social and cultural dynamics that influence interpretations of touch.

Researchers are experimenting with different types of sensors and materials to give robots a sense of touch and an ability to touch. Often multiple tactile sensors are used at once. In a survey of tactility for robots, Li and colleagues (2020) showed a NAO robot with hexagon-shaped mini-sensors designed to create a "multimodal artificial skin that provides vibration, temperature, force, and proximity information" (p. 1, 621). These sensors use a range of mechanisms—from electrical conductance and polarization to optical and magnetic—to detect tactile qualities of their environments and of things they are interacting with (Bartolozzi et al., 2016). While there is a growing range of tactile sensors (see Chapter Twelve), many of the most common tactile sensors embedded in social robots are those that sense pushing type forces but are unable to detect tactile signals related to holding objects. For instance, robots tend to struggle with gentle strokes or feeling the slippage of objects when grasping or stroking (Bartolozzi et al., 2016). For a robot, hugging a person is much easier than holding a slippery cup.

While tactility is important to helping a social robot interact with the world and to form connections with human users, the texturing of social robots is an equally important and intersecting concern. The feel of the robot—the way a human experiences its tactile engagement—is partially attributed to the texture of a device. While robots have typically been made of hard metals and plastics for durability and sometimes for hygienic reasons, research is increasingly focused on developing soft robotics, or social robots that have flexible, soft, and malleable skins and underlying forms. In some cases, this means covering a robot like NAO with a teddy bear outfit for experimental purposes (Burns et al., 2022) and in others it means developing bio-inspired skins and underlying forms made of soft, sometimes experimental materials (Hu & Hoffman, 2023). User communities also contribute to the texturing of devices by making outfits for them, dressing them in soft materials (Fig. 07.1.). For instance, the clothing and accessories forum for Living AI's EMO desktop companion is one of the most popular on its website and Living AI caters to this community by selling clothes to dress the EMO pet. Dressing social robots may serve the purpose of making them cuter or more stylized, but arguably these soft materials also play a role in the way owners experience the tactile qualities of robots like EMO (see also Chapter Six).

Several recent studies have provided experimental confirmation concerning the role texture plays in tactile perceptions of robots (Shiomi et al., 2020). One

Figure 07.1: Living AI's EMO in bear robe designed by Ara Archer Mira (source: Jason Archer).

study found that when robots had textured bodies made with materials like fur or soft silicone rubber, people rated them as more likable than robots with hard bodies (Yamashita et al., 2019). Researchers have also found that users prefer the soft, warm hug of a robot over a cold hug (Block & Kuchenbecker, 2019). This has prompted researchers to consider how to imbue social robots with thermoception and ways to physically express temperature variations. These developments add another technical layer to the touch systems developed for social robots and alter how texture and tactility are important technical and social components of robots. Touch is a multimodal system, made up of multiple intersecting and overlapping biological sensors and signals. It is further complicated by experiences of touch being interwoven with other sensory modalities. A preference for a warm hug over a cold hug, for instance, locates the issue of touch not only in the pressure felt but also in the temperature experienced. The point here, from a technical point of view, is that augmenting one dimension of texture or tactility may influence others because touch is always multimodal—and that is before considering the complexity introduced by psychological, social, and cultural dimensions that influence preferences, perceptions, and experiences of touch.

The science of human-robot interaction gives several reasons we should care about the texture and tactility of robots. Robot-initiated touch may influence social perceptions. Experimental evidence suggests when a cooperating robot initiates physical contact (versus, for instance, a human reaching out to the robot), that can evoke both positive and negative responses in the workplace (Arnold & Scheutz, 2018), influence whether people find a robot to be dependable (Cramer et al., 2009), and have potential persuasive powers. In one study, researchers found that participants who were touched by SoftBank Robotics' NAO robot during an emulated counseling session were more likely to comply with requests from the robot than those who were not touched (Hoffman & Krämer, 2021). Robotic touch shows potential to reduce perceptions of pain and alleviate stress (Geva et al., 2022), and some researchers argue that developing more robust touch systems for social robots could be important in using robots as therapeutic companions for children with autism (Burns et al., 2021).

As research on the tactility and texture of social robots continues to evolve, additional questions are being asked about the impact of robotic touch on social experiences like anthropomorphism, trust, companionship, and deception. Further, touch can mean different things in different contexts, so it is increasingly important to consider the social functions of robotic touch as distinct between companion robots and socially assistive robots (Paterson, 2023b). This additional research trajectory suggests that HRI researchers could benefit from incorporating theory and empirical evidence from nonverbal communication research and that our conceptualization of nonverbal communication can be expanded by considering how it functions between humans and robots. For instance, nonverbal communication research helps explain the social functions of touch, like the way that touch is used to initiate, regulate, and maintain the flow of communication in different relationships in different contexts. Touching hands may function to maintain a connection between romantic partners but would likely serve an initiating or regulating function between a patient and a doctor. Understanding these functions should inform the development of touch for robots if the intent is to design systems that are adaptable to various contexts and aim to achieve different goals in appropriate ways. And how might touch function as a form of nonverbal communication differently in human-robot relationships? More generally, human-human touch is still underexplored and to develop robotic systems that are accessible, inclusive, and inviting to a range of users, scholars will need to do more to understand the relationship between touch, perception, and emotion (Paterson, 2023a). In all cases, it is increasingly clear that robot touch matters and there is much work to be done in the area.

CHALLENGING THE *EMBRACE* ORIENTATION OF MACHINE TOUCH

There is a need for more research that considers the social, emotional, relational, and health-related impacts of social robotic touch systems—what could be termed the *embrace* orientation. But there is also a need to ask questions that get beyond treating touch in normative ways. Normative approaches encourage more prosocial relationships between humans and machines, make social robots more generally likable and make it easier for them to navigate their everyday environments and to do specific tasks. Normative approaches, in other words, operate from the assumption that social robots are beneficial to individuals and to society, if we can just get the design right. These are important considerations, but they seem rooted in early research on the Huggable which arguably promoted a paradigm around touch as functional and emotional while mostly omitting critical and cultural perspectives.[2]

Critical and cultural perspectives on touch and robots challenge us to question why research develops the way it does, to ask why design choices are made and consider who they benefit, and to examine the "social, cultural, and political challenges" that arise with the generation and dissemination of robots into society (Iliadis, 2023, p. 118). The perspectives treat technologies, like robot touch, as constructed through sociotechnical articulations, meaning the tactility and texture of robots is not just a set of technical features that arise from advances in technical knowledge and the development of new materials. The development and experience of certain forms of tactility and texture is, instead, also a set of social and political negotiations that are influenced by, express, and reinforce cultural beliefs and desires. To that end, the following sections lay out provocations related to care, labor, gender, and power. The provocations are meant to generate critical questions to aid researchers and developers, and to raise awareness about the societal stakes of robot touch, especially as the impacts will be felt not just by those designing or researching these robots but by the general public.

Taking Huggable related explorations as an example, scholars should question the assumptions informing the research, its findings, and its intersections with subsequent research in the field. Research making claims about the affective, health, and social implications of touch and social robotics predominately emerges from the lab (an isolated and controlled environment) and has yet to be grounded or explored in everyday contexts. At its best, lab-based research

---

2 Critical and cultural perspectives involving robot touch are starting to gain more traction in the field of HRI, but mostly remain at the margins (see Bucci et al., 2023; Barker & Jewitt, 2023; Paterson et al., 2023 for examples).

attempts to account for the complexity and multiplicity of human-robot interaction, as exemplified by recent robot-hug research, which suggests a set of design considerations based on multiple ways of touching and on helping robots perceive when humans do not want to be touched (Block et al., 2023). But even the best lab research cannot account for interactions in the wild, nor can it account for larger social implications. This creates both a set of limitations and a set of opportunities. In health contexts, how do definitions of affective robotic touch become associated with care? What are the implications of having so much research in this area focused on how to establish an emotional connection between humans and robots via touch? Whose affective states are being translated and excluded? How are the benefits and harms distributed when touch is used to create more intimate relationships? How are issues of unwanted or abusive touch considered when developing tactile and textured systems for machines? How can we account for the cultural and social contexts that may influence the ways touch is felt and understood between humans and machines—or stated another way, "what kind of touch is produced when we are unaware of the needs and desires of that what/whom we are reaching for?" (Puig de la Bellacasa, 2009, p. 300). What do people want huggable machines to feel like and how do they want them to hug, if they want to hug them at all?

Questions about the tactility and texture of social robots could also venture into new contexts where robots are already present by expanding notions of social touch beyond care and intimacy. Manufacturing represents one area where robots are already firmly embedded but is also a space where robots are rarely considered social unless they are visually anthropomorphized or use voice for interaction. Considerations about tactility and texture align with how to get a job done rather than how the machine influences human co-workers, alters the nature of work, and impacts the cultural environment of the workplace. Barker and Jewitt (2022; 2023) provide illustrative explorations that consider how the work of touch takes on new configurations with the introduction of tactile robots into workspaces, altering who is allowed to perform tactile work that is constructed as dirty and dangerous.

Researchers interested in gender, touch, and robots could also focus on media representations which often reinforce the gendered stereotypes that attribute soft skin with the feminine and metallic bodies with the masculine. Across a spectrum of films depicting social robots, the associations are surprisingly robust. Female-presenting robots are almost always skinned or rounded to create the illusion of softness (e.g., Ava in *Ex Machina* [2014], M3GAN in *M3GAN* [2022], EVE in *WALL-E* [2008]). Robots presented as metal skeletons without skin are typically depicted as male, signified by their names, voices, or other normative mannerisms (e.g., Jeff in *Finch* [2017], Chappie in *CHAPPIE* [2015], WALL-E in *WALL-E* [2008]). Looking toward media representations of the tactility and texture of fictional robots may reveal associations between touch and gender presentations that influence the development of touch systems for everyday robotics. At the very least, tracing the influences and tensions that cut across representations in the media, developments in the lab, and everyday encounters provide a critical dimension to consider the tactility and texture of social robots and their potential implications. At the intersection of media and inquiry, how do media representations of robot touch influence the development of robots and reinforce or challenge gender stereotypes? How do representations and research that equate soft robotics with affective touch, care, and health intersect and reinforce metaphors about softness and empathy which are historically coded female? Is there a role for tactility and texture designers to play in developing robot touch in ways that consciously transgress gender norms and promote greater inclusivity?

Taking inspiration from analogs in machine vision and listening research represents one other way to move beyond the dominant paradigm of touch in social robotics. Books tracing the historical, cultural, and social construction of machine vision systems articulate the ways vision is more than a set of technical features and perceptual capacities, indicating how they also express the cultural logics of developers who design them and shape the ways humans and non- humans see the world (Dobson, 2023; Rettberg, 2023). Thinkers who are investigating sociotechnical constructions of machine listening illustrate the stakes of operationalizing *listening* for machines—stakes that are connected to but go well beyond dyadic conceptualizations of human-robot interaction. The stakes include altering how individuals comport themselves when they think a machine is listening, extending state surveillance and corporate power through the implementation and control of listening machines and the data they produce, and turning social interaction into an extractable resource for commercialization (e.g., listening for keywords to deploy targeted ads; Napolitano & Grieco, 2021; Sterne, 2022). As Sterne (2022) succinctly states, "machine listening never 'just' listens" (p. 3) because machines that listen are meant to listen for a reason. Likewise, machine touch never "just" touches. More than a slogan, it is a call to action. Operationalizing touch for robots is a technical, social, and political process that requires us to ask important questions like: What constitutes robot touch and who decides? How might robot touch influence human touch and emotional connections between humans? How could robot touch be used to discriminate? What happens to the touch data collected in interactions and who controls it? How could governments or corporations exploit emotional connections generated through robot touch to concentrate power?

Answering the questions generated in these provocations presents a challenge to the *embrace* orientation, one that assumes developing robot touch to enable deeper emotional and social connections is a positive pursuit. These pursuits may arise from the best intentions, as with research focused on developing robot touch to promote health, but without a myriad of diverse perspectives at the table, well-intentioned research may lead to designs that harm as much as they heal. For instance, how do we ensure that robot touch does not become a kind of torture for people with diseases like fibromyalgia or who have social phobias related to histories of physical abuse? How do we ensure that assumptions about a desire to be touched, typically associated with females, are not unquestionably forwarded as research and design continue? Challenging the *embrace* orientation means asking hard questions about when, where, how, and why we may or may not want robot touch, what forms it should take, what functions it should enable, and who it is meant to touch or whose touch it is meant to replace. Challenging the *embrace* orientation is about bringing in diverse perspectives to create more inclusive forms of robot touch, including the right to not be touched, anticipating abuses, and mitigating potential harms.

If machine touch never "just" touches, we must consider the bodies and spaces that could be impacted. We must include potentially impacted communities in the decision-making process when developing and designing the tactile and texture features that constitute the technical aspects of robot touch. And we must consider the ways that these features could be exploited to gain control and concentrate power. In a world barreling toward greater autonomy for social robots we must think carefully about questions of tactility and texture because endowing our robots with touch is endowing them with power—power to heal and harm, to comfort and violate, to create and destroy. If machine touch never "just" touches, we must proceed gently.

REFERENCES

Arnold, T., & Scheutz, M. (2018). Observing robot touch in context: How does touch and attitude affect perceptions of a robot's social qualities? In *Proceedings of the International Conference on Human-Robot Interaction* (pp. 352–360). ACM.

Ballarini, G. (2003). Pet therapy. Animals in human therapy. *Acta Bio Medica*, 74, 97–100.

Barker, N., & Jewitt, C. (2022). Filtering touch: An ethnography of dirt, danger, and industrial robots. *Journal of Contemporary Ethnography*, 51(1), 103–130.

Barker, N., & Jewitt, C. (2023). Collaborative robots and tangled passages of tactile-affects. *ACM Transactions on Human-Robot Interaction*, 12(2), no. 19.

Bartolozzi, C., Natale, L., Nori, F., & Metta, G. (2016). Robots with a sense of touch. *Nature Materials*, 15(9), 921–925.

Bucci, P., Marino, D., & Beschastnikh, I. (2023). Affective robots need therapy. *ACM Transactions on Human-Robot Interaction*, 12(2), 1–22.

Block, A. E., & Kuchenbecker, K. J. (2019). Softness, warmth, and responsiveness improve robot hugs. *International Journal of Social Robotics*, 11(1), 49–64.

Block, A. E., Seifi, H., Hilliges, O., Gassert, R., & Kuchenbecker, K. J. (2023). In the arms of a robot: Designing autonomous hugging robots with intra-hug gestures. *ACM Transactions on Human-Robot Interaction, 12*(2), no. 18.

Burns, R. B., Seifi, H., Lee, H., & Kuchenbecker, K. J. (2021). Getting in touch with children with autism: Specialist guidelines for a touch-perceiving robot. *Paladyn, Journal of Behavioral Robotics, 12*(1), 115–135.

Burns, R. B., Lee, H., Seifi, H., Faulkner, R., & Kuchenbecker, K. J. (2022). Endowing a NAO robot with practical social-touch perception. *Frontiers in Robotics and AI, 9*, 840335.

Cramer, H., Kemper, N., Amin, A., Wielinga, B., & Evers, V. (2009). "Give me a hug": The effects of touch and autonomy on people's responses to embodied social agents. *Computer Animation and Virtual Worlds, 20*(2–3), 437–445.

Dobson, J. E. (2023). *The birth of computer vision.* University of Minnesota Press.

Field, T. M. (2001). *Touch.* MIT Press.

Geva, N., Hermoni, N., & Levy-Tzedek, S. (2022). Interaction matters: The effect of touching the social robot PARO on pain and stress is stronger when turned ON vs. OFF. *Frontiers in Robotics and AI, 9*, 926185.

Hoffmann, L., & Krämer, N. C. (2021). The persuasive power of robot touch. Behavioral and evaluative consequences of non-functional touch from a robot. *PLOS ONE, 16*(5), e0249554.

Hu, Y., & Hoffman, G. (2023). What can a robot's skin be? Designing texture-changing skin for human–robot social interaction. *ACM Transactions on Human-Robot Interaction, 12*(2), no. 26.

Iliadis, A. (2023). Critical and cultural approaches to human-machine communication. In A. L. Guzman, R. McEwen, & S. Jones (Eds.), *The SAGE Handbook of Human-Machine Communication* (pp. 117–126). SAGE.

Li, Q., Kroemer, O., Su, Z., Veiga, F. F., Kaboli, M., & Ritter, H. J. (2020). A review of tactile information: Perception and action through touch. *IEEE Transactions on Robotics, 36*(6), 1619–1634.

Napolitano, D., & Grieco, R. (2021). The folded space of machine listening. *SoundEffects—An Interdisciplinary Journal of Sound and Sound Experience, 10*(1), 173–189.

Paterson, M., Hoffman, G., & Zheng, C. Y. (2023). Introduction to the special issue on "Designing the robot body: Critical perspectives on affective embodied interaction." *ACM Transactions on Human-Robot Interaction, 12*(2), no. 14.

Paterson, M. (2023a). Inviting robot touch (by design). *ACM Transactions on Human-Robot Interaction, 12*(2), no. 16.

Paterson, M. (2023b). Social robots and the futures of affective touch. *The Senses and Society, 18*(2), 110–125.

Pawling, R., Cannon, P. R., McGlone, F. P., & Walker, S. C. (2017). C-tactile afferent stimulating touch carries a positive affective value. *PLOS ONE, 12*(2), e0173457.

Puig de la Bellacasa, M. (2009). Touching technologies, touching visions. The reclaiming of sensorial experience and the politics of speculative thinking. *Subjectivity, 28*(1), 297–315.

Rettberg, J. W. (2023). *Machine vision: How algorithms are changing the way we see the world.* Polity.

Schirmer, A., Croy, I., & Ackerley, R. (2023). What are C-tactile afferents and how do they relate to "affective touch"? *Neuroscience & Biobehavioral Reviews, 151*, 105236.

Shiomi, M., Sumioka, H., & Ishiguro, H. (2020). Survey of social touch interaction between humans and robots. *Journal of Robotics and Mechatronics, 32*(1), 128–135.

Sterne, J. (2022). Is machine listening listening? *Communication+1, 9*(1), 1–4.

Stiehl, W. D., Lieberman, J., Breazeal, C., Basel, L., Lalla, L., & Wolf, M. (2005). Design of a therapeutic robotic companion for relational, affective touch. *IEEE International Workshop on Robot and Human Interactive Communication* (pp. 408–415). IEEE.

Stiehl, W. D., & Breazeal, C. (2005). Affective touch for robotic companions. In J. Tao, T. Tan, & R. W. Picard (Eds.), *Affective computing and intelligent interaction* (pp. 747–754). Springer.

Yamashita, Y., Ishihara, H., Ikeda, T., & Asada, M. (2019). Investigation of causal relationship between touch sensations of robots and personality impressions by path analysis. *International Journal of Social Robotics, 11*(1), 141–150.

CHAPTER EIGHT

Gesture & Posture: Conveying Simulated Emotional States

SARA ALI

Stepping into the world where silence speaks volumes, where every movement narrates a tale, and where the language of gestures and postures paints the vivid emotions of our beloved robotic companions like R2D2—a famous fictional robot from Star Wars. R2-D2 is able to communicate through a series of beeps, whistles, and movements of its dome-shaped head and various appendages. For example: when excited or alarmed, R2-D2 might rapidly rotate its dome head. When expressing concern, it might tilt slightly or emit a lower-toned series of beeps. To indicate urgency, it might bob up and down or move rapidly from side to side. Similarly, when feeling content or satisfied, it might emit a series of cheerful beeps while standing tall and steady. Outside of the strong personalities in fiction, a real-world example is the dog-like Spot from Boston Dynamics, whose bodily movements can be interpreted as communication. For example: It may seem that Spot is expressing interest or curiosity when it moves its head and sensor panels to focus on something. It may seem cautious or unsure when it slows down its movements or pauses briefly. Similarly, it may seem as though Spot has particular intentions or desires when it uses its arm to interact with objects.

Although robots lack the innate ability to express emotions or intentions in the same way humans do, engineers are working to imbue them with nonverbal communication capabilities, or ways of sending messages using only their bodies—and not words. Nonverbal cues include facial expressions, gestures and movements, visual and audio cues, sensors, and feedback. Key among these forms are body language. Kinesic communication is a formal term for body language, which involves communication through body movements (Zabala et al., 2021).

The word "kinesic" derives from "kinesis" meaning "movement" which refers to hand, arm, body, and head actions (Hans & Hans, 2015). Combinations of nonverbal cues—distinct from but often congruent with its speech—enable a diverse range of behaviors for in support of its sociability (Knight, 2011; Ritschel et al., 2020).

In particular, robots' abilities to perform gestures and hold postures help to establish their social roles and agentic presence in social situations. As forms of non-verbal communication, gestures and postures are related but differ in their focus and expression. Gestures are the movement of a robot's limbs or body parts that are used for emphasizing points, giving directions, or expressing emotions and intentions in a dynamic and explicit manner. They tend to be more transient and may change frequently during interactions (Lhommet & Marsella, 2015). Posture, in turn, pertains to the overall alignment and position of a robot's body or parts over a longer duration representing robot's general stance, orientation, and demeanor throughout an interaction; posture conveys information about overall personality traits and interpersonal relations (Lee, 2020). Postures give subtle information about robot's expressed mood, attentiveness, or readiness for interaction. While gestures are typically more specific and context-dependent, posture provides a broader context to the robot's internal state.

THE MECHANICS OF GESTURE AND POSTURE

Both robot gestures and postures are governed by the shape of its body and the parameters for its movement. Body shape, or *morphology*, can be understood to be anthropomorphic, zoomorphic, or mechanomorphic. Anthropomorphic robots look like humans, in literal or essential terms, as they take up the whole or the gist of a humanoid body; these include robots like KASPAR (Kim et al., 2013) and NAO (Fujimoto et al., 2011). Zoomorphic robots look like animals, such the duckling-shaped as "Keepon" which is used for positive social interaction in therapies (Lee et al., 2012). Mechanomorphic or mechanoid robots have characteristically machine-like bodies that do not necessarily represent living bodies (see also Chapter One); for instance, the IGUS ReBeL robot is a mechanical arm with multiple segments and a single base that attaches to an anchoring surface (Sial et al., 2016).

Those body shapes in part govern what *parameters* a robot's body can have. Parameters are specifications that govern the mechanics of movement. For instance, a robot body with a largely static body (such as the spherical Leka robot; Cano et al., 2023) cannot engage parameters associated with arms or head movement, as can a robot with relevant joints (such as the learning robot Moxie; Hurst et al., 2020). There are multiple parameters that help to shape the exact gestures and postures a robot may take up, and of particular importance are those parameters that control motion: Velocity, acceleration, jerk, and curvature, each with substantial influence on seeing a robot as social (Saerbeck & Bartneck, 2010). Velocity is the speed and direction of a moving object, and in robotics is engaged in creating paths for things like robots or machines to move, using speed patterns to make sure the motions are smooth; this helps prevent putting too much strain on the motor and ensures everything moves more comfortably (García-Martínez et al., 2019), thereby influencing how a robot may keep pace with human interactions. Acceleration is defined as the change in velocity over a period of time (Ben-Ari & Mondada, 2017) influencing how smoothly or abruptly a robot transitions between different positions or gestures; controlled acceleration is important for accurately hitting specific poses or targets in a task. Jerk is the rate of change in acceleration over time, which is considered as an important factor for reducing vibrations and achieving accuracy for smoother motions (Sharkawy, 2021).

Among these, jerk in particular is central to movements that may leave impressions on humans. Smoother movements (i.e., lower jerk) helps the robot avoid overshooting or making sudden, uncoordinated movements. Moreover, jerk can measure how quickly acceleration changes. While jerk is less commonly used directly in robot programming, it indirectly affects the robot's motion quality. Minimizing jerk ensures that the robot's movements are comfortable to observe and interact with and hence constitutes naturalness; abrupt changes in acceleration (high jerk) can make the robot's motions appear unnatural or unsettling. Moreover, in terms of safety, sudden changes in jerk can lead to instability or accidents, which is especially important in applications where safety is a concern.

GESTURE AND POSTURE AS CARRIERS OF STATE INFORMATION

The rise of humanoid robots has motivated researchers to delve into and refine body language expression within robotic systems. Body language among humans often harnesses gestures and postures to communicate information about the emotions and thoughts of the sender, often revealing as much detail as verbal communication. Although robots do not (yet) experience emotion in the same way as humans (Hieida & Nagai, 2022), people may nonetheless *interpret* robots as having emotions or emotion-like internal states as a function of their behavior. For instance, the speed of a robot's vertical oscillation affects the quality of emotion being expressed in terms of arousal (Mahzoon et al., 2022). Even in the absence of intricate facial features, simulated emotions can be expressed with the help of movements of head, torso, legs, and arms that are used to link body movements and gestures (Ali et al., 2023). Human interpretation of emotions relies in part on movement characteristics (Roether et al., 2009), so the manipulation of a robot's movement parameters allows for the deliberate alteration of the robot's simulated emotional expressions (Hauser, 2013).

Engineers have developed a technique for communicating emotional behavior of robot using movement of its joints by varying parameters (Sial et al., 2016). It leverages features like speed, frequency, and joint angles to map these bodily actions to promote the interpretation of specific emotions—not only primary emotional categories e.g., anger, disgust, fear, happiness, sadness, and surprise (Ekman, 1992) but also more complex emotions like shame or hope. It's useful to draw from models of human emotions to consider how different gestural and postural parameters may be combined to convey emotional states. Consider, for instance, the Valence-Arousal-Dominance (VAD; Bălan et al., 2020) model of emotion. Valence, representing the spectrum of displeasure to joy, conveys favorable or unfavorable sentiment toward a subject. Arousal is the degree of emotional activation, spanning from sleep to excitement. Dominance indicates the extent to which a person has control over the emotional state, ranging from submissive to dominant (Bălan et al., 2020). Using this mapping of emotions with pleasure, arousal, and dominance, researchers have developed methods to presents the emotion using body gestures without any intricate facial features for communication, as discussed earlier (Ali et al., 2023a). Moreover, by considering the multidimensional aspects of emotions, they provide a more comprehensive understanding of the complexities inherent in human emotional states (Russell, 1980).

COMMUNICATING (ARTIFICIAL) EMOTION IN HUMAN-ROBOT INTERACTIONS

In part informed by those body shapes, gestures not only provide vital information within a message but also play an active role in shaping how we think and communicate as discussed in previous section, aiding in the process of message encoding and decoding (Clough & Duff, 2020). One such mechanism is known as an "affective loop" (Höök, 2009), encompassing an interactive procedure in which a human initially conveys emotions, often through bodily actions like gestures. Subsequently, the system (here, a robot) responds by generating their own gestural expressions of artificial emotion. That expression fosters some corresponding human response that gradually increases in the human's engagement with the system.

Over the past decade, there has been a significant surge in the development of robotic models of artificial emotions that can function in such affective loops.

These advances coincide with the widespread availability of commercial "affective" toys and robots (see Chapter Ten). Affective computing, sometimes referred to as artificial emotional intelligence, encompasses computing processes that are connected to, emerge from, or intentionally impact emotions and other affective experiences in communication (Picard, 1997). These technologies may go beyond merely recognizing human emotion to an enhanced capacity to interpret and intelligently respond to human emotions. These capacities include conveying their own artificial emotions to other agents communicating with them in same environment (Martínez-Miranda & Aldea, 2004). These internal processes can be projected outwardly in ways that humans can interpret, and those expressed emotions do indeed influence the social environment and interactions (Keltner & Gross, 1999), as body movements, postures, and gestures are instrumental in conveying emotional information (Clay et al., 2007). One example of these affective expressions is the case of research being carried out on Pepper robot to develop a system for natural talking gesture generation behavior (Zabala et al., 2022). Pepper is a semi-humanoid robot designed with the ability to read human emotions. This robot has been designed to automatically generate body gestures based on the sentiments in its speech to ensure a seamless synchronization when executed alongside speech (Zabala et al., 2022).Effective expression can be is accomplished through a combination of functional modules depending upon the scenarios. It can include modules such as a text-analyzer module, which involves categorizing or assigning predefined labels to textual data so that it can be automatically analyzed and categorized into specific classes or categories based on its content. This process is crucial for various applications, including sentiment analysis, spam detection, topic categorization etc. Similalrly, an emotion-selector module can provide an emotion value, which can be represented by gestures using varying speed and acceleration. Picking up from emotion selector module, a gestures-generator module can be responsible for crafting speaking gestures based on normative gestures that humans might make in the same situation. These gestures are meticulously generated to ensure a seamless synchronization when executed alongside speech. During this representation, the speed of the gestures is adjusted through the velocity parameter based on the intended emotion (Ali et al., 2023). For instance, if the emotion is interpreted as positively valanced, a gesture will be executed with liveliness and swiftness reflecting the upbeat nature of the emotion. Conversely, if the emotion is characterized as negatively valanced, a gesture will be executed at a slower pace aligning with the subdued nature of that particular emotion.

However, the conveyance and interpretation of emotion is not so simple as the norms associated with human expression do not always translate well to robots. This means that expression of artificial emotions is sometimes misinterpreted. For instance, for one social robot designed to express just eight emotions, human observers were able to recognize only five kinds of emotional body language correctly about 55% of the time (McColl & Nejat,2014). Moreover, even specific emotions can have complex variations. For instance, negative, high-arousal emotions like anger may require specialized engineering to express and to correspond with anger-based utterances with choice of the gesture accompanying such utterances (Ajibo et al., 2020). Because robots have complex bodies (as do humans) and the meaning of a gesture can change based on context, the expressive functions of robot gestures must also be considered in relation to facial poses and expressions (Ju & Kang, 2012). In short, the challenge of creating multimodal, multidimensional, and multi-contextual expressions of emotions by robots is key to how bodily non-verbal communication help to establish a social relationship with humans (Fong et al., 2003; Zabala et al., 2021).

BEYOND HUMAN-LIKE EMOTION?

I have argued that robot posture and gesture parameters for shape and movement, respectively, are key to how robots may foster interpretations of emotional expression. For specific tasks and expressions, robot programmers and designers can consider these factors to create motion profiles that align with the intended communication or function also such as expressing happiness, showing understanding, providing instructions, and displaying urgency. However, these dynamics must be crafted with consideration for context which includes the choice of velocity, acceleration, and jerk profiles should align with the specific situation, cultural norms, and the desired emotional or functional impact. Robots with the ability to modulate these parameters effectively can engage users more naturally and convey their gestures and postures with greater clarity and expressiveness. However, there are important considerations around the assumption that more human-like emotional expressions will be more effective in human-robot interactions.

First, despite the emphasis on bodily gestures and postures in this chapter, there is relatively limited understanding of how to convey nuanced emotions using a robot's full body gestures. Instead, as with the robot Ameca (Engineered Arts, n.d.) or Sophia (Hanson Robotics, n.d.), there is often an emphasis on facial expressions. Complex facial expressions could indeed be especially important to advancing robot sociality (see Chapter Two), however a dependency on intricate facial features adds to a robot's cost and maintenance requirements—making it infeasible or impractical for many social contexts. As such, it is critical to work toward generating believable expressions of artificial emotion through more core body postures and gestures.

Second, there is an assumption that more anthropomorphic (i.e., more human-like) body formations are better. However, per Mori's Uncanny Valley hypothesis, one's comfort with an agent is directly proportional to the agent's human-likeness—but only to a point where it becomes not-quite-human, inducing feelings of discomfort. This discomfort may emerge from seeing anthropomorphic technologies as strangely imperfect (Alvarez Perez et al., 2020) or even menacing (Ferrari et al., 2016; Stein et al., 2019), evoking aversion at both cognitive and emotional levels. This aversion can manifest through phenomena such as cognitive dissonance, thoughts of mortality, or concerns about the erosion of human uniqueness (Złotowski et al., 2017). In other words, relying on highly human forms of emotion expression could risk alienating humans by creating a sense of unease.

If we allow these considerations to temper our initiatives around facial and anthropomorphic expressions of emotion, we position ourselves to be open to other possibilities for how robot bodies and body movements might express artificial emotion. As robots become more mechanically sophisticated, is it possible that they may engage in a more nuanced use of velocity, acceleration, and jerk that go beyond mimicking human behavior but potentially shifts to instead convey its machinic internal states? Consider the possibilities: As robots harness speed and tempo in their gestural movements, could they transcend merely simulating human emotions and instead achieve a heightened level of expression that represents their inner workings? We should consider what might arise from this heightened expressiveness—perhaps an amplified intensity, serenity, or urgency that helps us to understand them better? Although that sounds beneficial, it is also possible that humans might find such expression challenging due to the unfamiliarity of the signals or the rapidity with which they are conveyed, thereby struggling to relate to or comprehend these expressions.

Another consideration is whether future robots could master transitions between positions and gestures to an extent where their fluidity exceeds the aesthetics and naturalness of human motion? Could advancements lead to robots surpassing human standards of safety in motion, ensuring stability, and preventing accidents in ways that humans cannot guarantee? In such a zeitgeist, we might no longer see humans as a gold standard for expression, but instead seek to create machines that realize their own potentials for machinic expression—and so craft a new paradigm in how we perceive, interpret, and engage with non-verbal communication in the realm of artificial emotional intelligence.

REFERENCES

Ajibo, C. A., Ishi, C. T., Mikata, R., Liu, C., & Ishiguro, H. (2020). Analysis of body gestures in anger expression and evaluation in android robot. *Advanced Robotics, 34*(24), 1581–1590.

Ali, S., Mehmood, F., Iqbal, K. F., Ayaz, Y., Sajid, M., Sial, M. B., Malik, M. F., & Javed, K. (2023). Human robot interaction: Identifying resembling emotions using dynamic body gestures of robot. In *Proceedings of the International Conference on Artificial Intelligence* (pp. 39–44). IEEE.

Alvarez Perez, J., Garcia Goo, H., Sánchez Ramos, A., Contreras, V., & Strait, M. (2020). The uncanny valley manifests even with exposure to robots. In *Proceedings of the International Conference on Human-Robot Interaction* (pp. 101–103). ACM.

Bălan, O., Moise, G., Petrescu, L., Moldoveanu, A., Leordeanu, M., & Moldoveanu, F. (2020). Emotion classification based on biophysical signals and machine learning techniques. *Symmetry, 12*(1), no. 21.

Ben-Ari, M., & Mondada, F. (2017). Robots and their applications. In *Elements of robotics* (pp. 1–20). Springer.

Cano, S., Díaz-Arancibia, J., Arango-López, J., Libreros, J. E., & García, M. (2023). Design path for a social robot for emotional communication for children with Autism Spectrum Disorder (ASD). *Sensors, 23*(11), 5291.

Clay, A., Couture, N., & Nigay, L. (2007). Emotion capture based on body postures and movements [preprint]. <https://arxiv.org/abs/0710.0847>

Clough, S., & Duff, M. C. (2020). The role of gesture in communication and cognition: Implications for understanding and treating neurogenic communication disorders. *Frontiers in Human Neuroscience, 14*, 569053.

Ekman, P. (1992). Are there basic emotions? *Psychological Review, 99*(3), 550–553.

Engineered Arts. (n.d.). *Ameca*. <https://www.engineeredarts.co.uk/robot/ameca/>

Ferrari, F., Paladino, M. P., & Jetten, J. (2016). Blurring human–machine distinctions: Anthropomorphic appearance in social robots as a threat to human distinctiveness. *International Journal of Social Robotics, 8*(2), 287–302.

Fong, T., Nourbakhsh, I., & Dautenhahn, K. (2003). A survey of socially interactive robots. *Robotics and Autonomous Systems, 42*(3–4), 143–166.

Fujimoto, I., Matsumoto, T., de Silva, P. R. S., Kobayashi, M., & Higashi, M. (2011). Mimicking and evaluating human motion to improve the imitation skill of children with autism through a robot. *International Journal of Social Robotics, 3*(4), 349–357.

García-Martínez, J. R., Rodríguez-Reséndiz, J., & Cruz-Miguel, E. E. (2019). A new seven-segment profile algorithm for an open source architecture in a hybrid electronic platform. *Electronics, 8*(6), no. 652.

Hans, A., & Hans, E. (2015). Kinesics, haptics and proxemics: Aspects of non-verbal communication. *IOSR Journal of Humanities and Social Science, 20*, 47–52.

Hanson Robotics. (n.d.). *Sophia*. <https://www.hansonrobotics.com/sophia/>

Hauser, K. (2013). Recognition, prediction, and planning for assisted teleoperation of freeform tasks. *Robotics: Science and Systems, 8*(812), 121–128.

Hieida, C., & Nagai, T. (2022). Survey and perspective on social emotions in robotics. *Advanced Robotics, 36*, 17–32.

Höök, K. (2009). Affective loop experiences: Designing for interactional embodiment. *Philosophical Transactions of the Royal Society B: Biological Sciences, 364*(1535), 3585.

Hurst, N., Clabaugh, C., Baynes, R., Cohn, J., Mitroff, D., & Scherer, S. (2020). Social and emotional skills training with embodied Moxie [preprint]. <https://arxiv.org/abs/2004.12962v1>

Ju, M. H., & Kang, H. B. (2012). Emotional interaction with a robot using facial expressions, face pose and hand gestures. *International Journal of Advanced Robotic Systems, 9*(3).

Keltner, D., & Gross, J. J. (1999). Functional accounts of emotions. *Cognition and Emotion*, *13*(5), 467–480.

Kim, E. S., Berkovits, L. D., Bernier, E. P., Leyzberg, D., Shic, F., Paul, R., & Scassellati, B. (2013). Social robots as embedded reinforcers of social behavior in children with autism. *Journal of Autism and Developmental Disorders*, *43*(5), 1038–1049.

Knight, H. (2011). Eight lessons learned about non-verbal interactions through robot theater. In *Proceedings of the International Conference on Social Robotics* (pp. 42–51). Springer.

Lee, D. (2020). Gesture, posture, facial interfaces. In *Encyclopedia of Robotics* (pp. 1–10). Springer.

Lee, J., Takehashi, H., Nagai, C., Obinata, G., & Stefanov, D. (2012). Which robot features can stimulate better responses from children with Autism in robot-assisted therapy? *International Journal of Advanced Robotic Systems*, *9*(3).

Lhommet, M., & Marsella, S. C. (2015). Expressing emotion through posture and gesture. In *The Oxford handbook of affective computing* (pp. 273–285). Oxford University Press.

Mahzoon, H., Ueda, A., Yoshikawa, Y., & Ishiguro, H. (2022). Effect of robot's vertical body movement on its perceived emotion: A preliminary study on vertical oscillation and transition. *PlOS one*, *17*(8).

Martínez-Miranda, J., & Aldea, A. (2004). Emotions in human and artificial intelligence. *Computers in Human Behavior*, *21*(2), 323–341.

McColl, D., & Nejat, G. (2014). Recognizing emotional body language displayed by a human-like social robot. *International Journal of Social Robotics*, *6*(2), 261–280.

Picard, R.W. (1997). *Affective computing*. MIT Press.

Ritschel, H., Kiderle, T., Weber, K., Lingenfelser, F., Baur, T., & André, E. (2020). Multimodal joke generation and paralinguistic personalization for a socially-aware robot. In *Advances in practical applications of agents, multi-agent systems, and trustworthiness* (pp. 278–290). Springer.

Roether, C. L., Omlor, L., Christensen, A., & Giese, M. A. (2009). Critical features for the perception of emotion from gait. *Journal of Vision*, *9*(6), 1–32.

Russell, J. A. (1980). A circumplex model of affect. *Journal of Personality and Social Psychology*, *39*(6), 161–178.

Saerbeck, M., & Bartneck, C. (2010). Perception of affect elicited by robot motion. In *Proceedings of the International Conference on Human-Robot Interaction* (pp. 53–60). ACM

Sharkawy, A. (2021). Minimum jerk trajectory generation for straight and curved movements: Mathematical analysis [preprint]. <https://arxiv.org/abs/2102.07459>

Sial, S. B., Sial, M. B., Ayaz, Y., Shah, S. I. A., & Zivanovic, A. (2016). Interaction of robot with humans by communicating simulated emotional states through expressive movements. *Intelligent Service Robotics*, *9*, 231–255.

Stein, J.-P., Liebold, B., & Ohler, P. (2019). Stay back, clever thing! Linking situational control and human uniqueness concerns to the aversion against autonomous technology. *Computers in Human Behavior*, *95*, 73–82.

Zabala, U., Rodriguez, I., Martínez-Otzeta, J. M., & Lazkano, E. (2021). Expressing robot personality through talking body language. *Applied Sciences*, *11*(10), 4639.

Zabala, U., Rodriguez, I., Martínez-Otzeta, J. M., & Lazkano, E. (2022). Modeling and evaluating beat gestures for social robots. *Multimedia Tools and Applications*, *81*(3), 3421–3438.

Złotowski, J., Yogeeswaran, K., & Bartneck, C. (2017). Can we control it? Autonomous robots threaten human identity, uniqueness, safety, and resources. *International Journal of Human-Computer Studies*, *100*, 48–54.

CHAPTER NINE

Text & Speech: Robot Communication Design

KRISTINE L. NOWAK & ANTONIO CHELLA

Human-robot interaction studies examine what happens when robots and people communicate and how to best design the robot interface. A robot interface includes everything people see, touch, or hear as they engage with a robot, including the way they communicate. Social robots are increasing their communication skills, enabling them to generate, comprehend, and respond to a wider selection of messages or commands, and to do so more fluidly and in ways that are increasingly similar to human-human interaction. Engineering and design decisions determine both the robot's function and appearance, as well as the way people engage with the robot, which influences perceived capabilities and expectations (Banks, 2020; Rosenthal-von der Pütten et al., 2018). Thus, a major challenge in the design and engineering of social robots is balancing (a) the need to be agile in speech and behavior so they may adapt to individual humans and contexts, against (b) human requirements for affordability, fit for human social spaces and processes, and other context-driven functionality (Lin et al., 2017).

When a robot includes interface features designed to make interactions more natural or human-like, it is called anthropomorphic design (Fong et al., 2003; Rosenthal-von der Pütten & Kramer, 2014), which includes behaviors such as speech and speech patterns that allow robots to verbally interact in more human-like ways (Geraci et al., 2021). The goal of anthropomorphic design is to make it more intuitive, easier, and more enjoyable for people to communicate and engage with robots. Research suggests more anthropomorphic robots are generally more liked, deemed more credible, and interacted with as if they have human-like minds and abilities, though this is not true of all robots in all contexts, and the relationship is not precisely linear (Banks, 2021; de Visser et al., 2016), further complicating design decisions. While social robots can be designed to utilize many modes of communication including nonverbal behaviors, the most common modes are text in which the robot receives or shares written words on a screen, and speech whereby a robot can understand and transmit oral words and sentences. This chapter defines these two modes of social robot communication and examines related engineering challenges.

ENGINEERING AND DESIGN CHALLENGES OF SOCIAL ROBOT COMMUNICATION

As with any communicator, when robots communicate, they must have a message to share, a way to share it, and a receiver who can interpret and respond to the message. The traditional Mathematical Model (Shannon & Weaver, 1949) model that defines communication as a process of information transmission where a source generates meaning and creates a message, which is shared using a specific form or modality (e.g., text or speech) with a receiver. The receiver processes and translates the message to the intended recipient of the message. This process is interactive and requires both sender and receiver to overcome noise that may distort or interfere with the transmission, reception, or understanding of messages within the communication process. In these terms, the mathematical model can help us understand and predict communication between two humans as well communication between a human and a robot, because a successful interaction requires a source able to generate a message to share and a receiver capable of understanding and responding appropriately.

Labels and Creating Meaning

As described in the Mathematical Model above, to be a communicator, robots must be able to generate a message, which is not an insignificant engineering challenge. To generate a message, they must be able to process, learn, recognize, and understand labels and references people are using. Thus, they must be able to perceive and identify objects, sort them into categories, and then record and associate those objects with labels. For example, to be able to follow the command to open a door, a robot would have to know what a door is, know the location of the door, and understand what is meant by the "open" command.

To solve this challenge, social robots can be programmed to have a "perception" module and/or a "proprioception" module allowing them to identify objects and connect them to labels or referents. Typically, the cognitive cycle of a robot's architecture starts with the perception module, which is a software program that can analyze the signals coming from the external environment, such as the names and locations of objects, connect them to related objects, and place them in the robot's memory. For example, the robot can use a camera to see an object and assign or translate the object into a label, and store in its working memory (i.e., their "phonological store"), the label and related information about the object (such as its location or purpose). The label and information about the object are combined with other related words and items in the working memory, allowing the robot to connect similar items and rehearse and store them for later retrieval. Then, the robot may generate the phrase "I see an apple" and understand that the apple is fruit.

As robots learn words to label objects, the terms are added to their memory (phonological store) so they can recognize and respond to more diverse commands that match the accumulated words or labels. Innovations in processing capabilities have led to a wider range of terms and commands the robot can store, recognize, and use appropriately in social interactions. Robots with fewer learned labels are not able to fulfill certain commands or understand referents being made during an interaction, while more advanced robots have abilities to store more words and so are able to learn from their users; in such learning, so they can collect a wide array of commands and objects they recognize, allowing them to provide a wider selection of services across contexts (Alonso-Martín & Salichs, 2011). However, a robot who has not learned an object will perceive or "see" it but not be able to associate it with a label or respond to commands using the label. For example, if a robot has seen an object that it has previously learned is assigned the label "orange," it can respond to a command requesting them pick up the orange and associate it as part of the fruit category along with the apple. However, a robot who has not learned to associate the orange with the label "orange cannot respond appropriately to this command."

Translating Labels to Messages

Once robots have labels and words in their phonological store, they can be programmed to share information about labeled objects or respond to messages using the labels in a variety of ways, the most common of which are via written word (text) or by speaking and decoding or processing words orally (speech). Robots can receive messages either orally or textually. Reception of oral messages happens when they are able to apprehend and decode the sound waves of spoken word (the equivalent of human hearing or receiving the message). Reception of text messages happens when a robot receives a message via text that they accept and decode character-denotive data, as when it is sent through a keyboard that is connected to the robot. Depending on its designed capabilities, a robot may be able to understand oral communication but is only able to respond with text, or by some other combination of speech, text, and nonverbal communication. For example, a person could make a request for the robot to open a door either orally, by selecting an option from a touchscreen, or even by typing it onto a keyboard in text form (e.g., in scenarios with high noise when the robot is unable to hear or understand spoken language). The robot can be programmed to confirm understanding of the message by either bringing the item, with oral or text confirmation, or even with colored lights (Rosenthal-von der Pütten et al., 2018).

The integration and optimization of natural speech recognition systems into such real-world settings is one of the most significant challenges for social robot communication. Speech recognition systems are those in which robots utilize oral communication; hearing commands and responding with speech are the most "natural" because they require little learning or effort for people and are generally associated with anthropomorphic design. Although modern text-to-speech systems can generate near-perfect simulations of the human voice, there is an ongoing and complex challenge in developing a robotic voice and engagement style that best fits with the overall performance and goals of people engaging it (Ashok et al., 2022; Kühne et al., 2020).

SHARING ROBOT THOUGHTS AND INNER SPEECH

While communicating about the environment or objects is useful, there are other parts of social communication that are essential for humans to feel connected with another entity. For example, another advance in robot communication capabilities involves enabling it to share what is called "inner speech," or the robot's internal voice that can help people understand the robot's learning and decision- making process (Chella et al., 2020; Pipitone & Chella, 2021). When people engage in this inner-speech process, it is sometimes called *talking to themselves*, and can also be called private speech, internal monologue or dialogue, self- directed thought, subvocal or covert speech. During inner speech, people reflect on their options and choices as they make decisions, becoming consciously aware of options and resources, retrieve learned facts, and even learn and store new information as they seek to actively control and regulate their behavior (Alderson-Day & Fernyhough, 2015). Inner speech can take many forms, such as condensed (few words) or expanded (full sentences), and monolog (using "I") or dialogue (asking questions and answering them using both "I" and "you"), and it can be shared with others to enhance understanding or trust.

Robots can be designed to use a proprioception module to develop and utilize inner speech to help them explore, describe, and understand static and dynamic scenes; this can help improve decision-making and situational awareness. This connection and rehearsal of stored items allows the creation of what is called the robot inner ear, which is part of the proprioception module that facilitates retrieval of labels when the object is encountered in the future; doing so allows a robot to discuss and engage in discussions about these items in interactions (Pipitone & Chella, 2021). In this case, for example, when a robot hears the "I see an apple"—a phrase generated by itself by its inner ear—it activates the same computational processes that are activated when the robot effectively sees the object. The proprioception module in more advanced social robots analyzes the signals coming from the internal state of the robot as it is making sense of information and deciding on future action. When the robot generates inner speech, it orally communicates and shares with humans their inner monologue that is typically only presented to the self (Langland-Hassan & Vicente, 2018). Research continues to investigate whether sharing this inner monologue is desirable or helpful to the human-robot relationship, but preliminary evidence suggests it increases perceived credibility in many collaborative settings (Pipitone et al., 2021; Pipitone et al., 2023). The tradeoff is that the robot takes time to generate inner speech, which may frustrate a person who wants the robot to follow a specific command or is not interested in hearing the inner speech. This is why there are settings enabling people to turn off inner speech for times when a person wants the robot to perform its tasks efficiently and with as little delay as possible.

NOISE IN SOCIAL ROBOT COMMUNICATION

As with all communication, there can be either informational or aural noise that interfere with the robots' ability to receive or understand the message intended by the sender. One source of informational noise occurs because of insufficient learning or programming as described above, where the robot does not have a label for an object, has a different label for an object than the one being used (e.g., cup instead of glass), or is for one reason or another not able to understand the referent or label being used to describe something. There is also aural noise, ambient or otherwise, such as other people talking or car noises that interferes with the robots' ability to understand a person's oral communication.

Theoretically and in ideal conditions, oral speech communication is the easiest way to interact with a robot and most systems perform quite well in quiet, controlled environments, such as indoor office spaces with minimal (or at least predictable) noise or acoustic interference. However, these ideal conditions and controlled environments are not always possible, and robots frequently need to operate in uncontrolled, noisy environments. For instance, robots may function as greeters in crowded stores or hospitals with many people talking at once or even in outdoor settings with varying levels of ambient noise (e.g., from wind or traffic). Unfortunately, even advanced robot speech-recognition systems struggle to comprehend and respond to messages or commands in environments with unpredictable ambient noise. Also, robots need time to learn to process voices and accents different from the ones they hear regularly or those used to train them, which can also make it difficult for them to understand some oral commands and communication. This leads to less-than-optimal speech recognition and compromised speech performance that may require people and robots to raise their voices and repeatedly articulate commands, frequently leading to frustration and miscommunication (Alonso Martin et al., 2020).

One solution for managing these challenges with speech communication is for robots to utilize text-based communication as a backup, which can potentially mitigate some of these issues. Speech-recognition systems using text-based communication is relatively simple and less prone to error from an engineering perspective. However, having people typing text to a robot or reading messages from robots in words on a screen can be awkward, inefficient, and unnatural in practice. Text-based interactions can be more precise but are inherently slower and a less intuitive compared to voice-based communication. Additionally, a change of modality (from speech to text) can lead to frustration or confusion as it requires the user to adjust to a different interface and changes expectations for the interaction. Even so, it can sometimes be the best option to facilitate the interaction if the robot is unable to understand or be understood due to noise or other considerations.

Additionally, when speech-recognition performance is limited, as in the ambient noise environments considered above, this may reduce the flow of the conversation and may also compromise the robots' capability to accurately assess or respond to a person (Gasteiger et al., 2022), which may lead to frustrating and less satisfying experiences. This is similar to a person communicating with a personal assistant such as Alexa or Siri in a noisy environment where the assistant barely recognizes words, requiring the person to repeat the same phrase many times. In this case, people tend to adopt a loud voice hoping to be understood by the robot, which may lead to further miscommunication, distrust and frustration that may lead people to be less inclined to utilize or interact with the robot.

DEMONSTRATING DESIGN AND USABILITY CHALLENGES: THE CASE STUDY OF PEPPER

It is useful to consider the engineering and usability challenges around speech and text communication through the example of Pepper. Pepper is a commercially available, 4-foot-tall, interactive, mobile, humanoid social robot that can communicate both via text through a chest-mounted screen and orally by a synthesized voice created by Aldebaran (2023). It is one of the most common humanoid robots employed for commercial, research, and teaching human-robot interactions because of its expressive face, capability to move the head and arms, richness of sensors and motors, and relative ease in programming. In addition to textual and oral outer speech, it can also be programmed to orally share information about its internal state by sharing inner speech.

In an example scenario used in some studies (Geraci et al., 2021), Pepper enters a space with fruits and cutlery on a table and the person and robot are instructed to set the table together. As described above, the perception module senses objects such as an apple and generates labels, such as <apple>, <round>, and <green>, which are sent to the robot's working memory that the robot can use later when the robot generates inner speech phrases like <I see an apple> that re-enter the working memory and become new input. The label <apple> is then confronted with the items stored in the robot memory and connected to other related labels by association, such as <fruit> and <orange>. The robot has a symbolic knowledge base where facts related to the scenario are stored. Then, comparing the robot's knowledge base with the generated phrase, the robot deduces that an apple is round and red and that it is a fruit, and then a new phrase like <apple is a fruit> is generated, rehearsed, and shared as the inner speech of the robot. This phrase, in turn, generates other phrases like <orange is a fruit too> and <I put fruit in the red basket>. The inner speech system allows Pepper to share its process or reasoning with people, which can make it easier for them to understand the reasons behind its behaviors. For instance, verbalizing the logic above might help people understand that Pepper might put apples and oranges in the same basket because it sees them both as belonging to the fruit category.

Having access to this inner speech has been shown to make people feel more comfortable and engaged with Pepper, which helps people successfully complete collaborative tasks with the robot, such as placing plates and forks in the right place as they set the table. People seem to like it when Pepper shares inner speech; research shows they trust it more and perceive it to be more alive, more intelligent, and easier to work with than when the robot does not share inner speech (Pipitone et al., 2021).

OPPORTUNITIES AND DESIGN CHALLENGES OF SOCIAL ROBOT COMMUNICATION

Social robot communication—whether by voice or by text—provides many exciting opportunities and design challenges as robots integrate into more human spaces. For robots to take advantage of these opportunities, designers must balance a robot's communication style (ensuring a fit with people's expectations) against its innovative capabilities (allowing it to overcome noise and processing difficulties). The ability to allow choice for individual differences and preferences is itself a design challenge, as communication style is sometimes built into the robot along with the memory and other instructions.

Some robots who fluidly speak sentences with realistic and human-sounding voices will be impressive and appropriate for some tasks and venues, as with collaborative tasks. At the same time, those same fluid speaking abilities that are engaging and impressive in one context could be distracting or unnecessary for other people or tasks such a person who just wants the robot to open the door or to vacuum the floor. While it may be good for some people in some contexts for the robot to help establish a comfortable relationship by explaining what the robot is capable of, these behaviors could also lead to disappointment or overreliance on the robot. For example, when a robot speaks with specific vocal inflections that imply that expressions are jokes, sarcasm, or other non-literal language, a person could begin to expect that the robot is able to get when someone is telling a joke—and then become upset when the robot doesn't "get" that person's joke or indicate that it is funny. Even so, if a robot fails at detecting humor in human expressions, a person may better understand how the joke was interpreted and not be as disappointed if they are able to hear the robot's inner monologue.

Beyond creating human expectations, designing highly anthropomorphic robots who can communicate more naturally using oral speech is expensive in many ways. Giving them natural and human-like voices or sentence structures during social interactions is a design challenge that requires extensive resources including time and money, both for those creating and those purchasing and operating the robots. It also requires a lot of control over the environment as it is more sensitive to noise and other issues. Thus, designers should weigh the value of investment in more natural or anthropomorphic design and determine whether the same task can be accomplished with a simpler design. Importantly, making robots more social, anthropomorphic, or "natural" is not necessary, desirable, or even appropriate for all people or contexts (Goetz et al., 2003; Lohse et al., 2008). This presents the usability challenge that is also an ethics and an engineering challenge in that the best design for a social robot lies in the synthesis of the robot's capabilities with the input and output strategies that are best varying people and interaction goals.

REFERENCES

Aldebaran. (2023). *Pepper documentation.* <https://www.aldebaran.com/>

Alderson-Day, B., & Fernyhough, C. (2015). Inner speech: Development, cognitive functions, phenomenology, and neurobiology. *Psychological Bulletin, 141*(5), 931–965.

Alonso Martin, F., Malfaz, M., Castro-González, Á, Castillo, J. C., & Salichs, M. Á. (2020). Four-features evaluation of text to speech systems for three social robots. *Electronics, 9*(2), no. 267.

Alonso-Martín, F., & Salichs, M. A. (2011). Integration of a voice recognition system in a social robot. *Cybernetics and Systems, 42*(4), 215–245.

Ashok, K., Ashraf, M., Thimmia Raja, J., Hussain, M. Z., Singh, D. K., & Haldorai, A. (2022). Collaborative analysis of audio-visual speech synthesis with sensor measurements for regulating human–robot interaction. *International Journal of System Assurance Engineering and Management.*

Banks, J. (2020). Theory of mind in social robots: Replication of five established human tests. *International Journal of Social Robotics, 12*(2), 403–414.

Banks, J. (2021). From warranty voids to uprising advocacy: Human action and the perceived moral patiency of social robots. *Frontiers in Robotics and AI, 8*, 670503.

Chella, A., Pipitone, A., Morin, A., & Racy, F. (2020). Developing self-awareness in robots via inner speech. *Frontiers in Robotics and AI, 7.*

de Visser, E. J., Monfort, S. S., McKendrick, R., Smith, M. A. B., McKnight, P. E., Krueger, F., & Parasuraman, R. (2016). Almost human: Anthropomorphism increases trust resilience in cognitive agents. *Journal of Experimental Psychology: Applied, 22*(3), 331–349.

Fong, T., Nourbakhsh, I., & Dautenhahn, K. (2003). A survey of socially interactive robots. *Robotics and Autonomous Systems, 42*(3), 143–166.

Gasteiger, N., Lim, J., Hellou, M., MacDonald, B. A., & Ahn, H. S. (2022). A scoping review of the literature on prosodic elements related to emotional speech in human-robot interaction. *International Journal of Social Robotics.*

Geraci, A., D'Amico, A., Pipitone, A., Seidita, V., & Chella, A. (2021). Automation inner speech as an anthropomorphic feature affecting human trust: Current issues and future directions. *Frontiers in Robotics and AI, 8.*

Goetz, J., Kiesler, S., & Powers, A. (2003). Matching robot appearance and behavior to tasks to improve human-robot cooperation. In *Proceedings of the International Workshop on Robot and Human Interactive Communication* (pp. 55–60). IEEE.

Kühne, K., Fischer, M. H., & Zhou, Y. (2020). The human takes it all: Humanlike synthesized voices are perceived as less eerie and more likable. Evidence from a subjective ratings study. *Frontiers in Neurorobotics, 14.*

Langland-Hassan, P., & Vicente, A. (2018). In Langland-Hassan P., Vicente A. (Eds.), *Inner speech: New voices.* Oxford University Press.

Lin, P., Abney, K., & Jenkins, R. (Eds.). (2017). *Robot ethics 2.0: From autonomous cars to artificial intelligence.* Oxford University Press.

Lohse, M., Hegel, F., & Wrede, B. (2008). Domestic applications for social robots—An online survey on the influence of appearance and capabilities. *Journal of Physical Agents, 2*(2), 21–32.

Pipitone, A., & Chella, A. (2021). What robots want? Hearing the inner voice of a robot. *iScience, 24*(4), 1–15.

Pipitone, A., Geraci, A., D'Amico, A., Seidita, V., & Chella, A. (2023). Robot's inner speech effects on human trust and anthropomorphism. *International Journal of Social Robotics*.

Pipitone, A., Geraci, A., D'Amico, A., Seidita, V., & Chella, A. (2021). Robot's inner speech effects on trust and anthropomorphic cues in human-robot cooperation [preprint]. <http://arxiv.org/abs/2109.09388>

Rosenthal-von der Pütten, A. M., & Kramer, N. C. (2014). How design characteristics of robots determine evaluation and uncanny valley related responses. *Computers in Human Behavior, 36*, 422–439.

Rosenthal-von der Pütten, A. M., Krämer, N. C., & Herrmann, J. (2018). The effects of human-like and robot-specific affective nonverbal behavior on perception, emotion, and behavior. *International Journal of Social Robotics, 10*(5), 569–582.

Shannon, C. E., & Weaver, W. (1949). *The mathematical theory of communication*. University of Illinois Press.

CHAPTER TEN

Screens & Links: Playful Affordances of Future Friends

KATRIINA HELJAKKA

Toys are items meant to be played with, and play implies engagement of various kinds. Toys and play evolve and reflect technological change. Technology-augmented toys, such as smart toy robots, are emerging quickly in the toy market. The advent of technology-augmented toys, such as smart and Internet-connected toys, produce new venues for play, which creates a new environment for the psychological study of play (Bergen, 2015). This chapter considers smart toy robots from two viewpoints—considerations of screens and internet-connective linkages, and considerations of how these dimensions of the toy robot components cater to playful interaction. Moreover, the chapter explores the potential of smart toy robots to be future friends by examining their screen-based and networked affordances through speculative toy fiction.

Fernaeus and colleagues (2010) characterize robots as active, tangible artifacts that interact directly with the world around them. Smart toy robots are *teachable machines* (Druga et al., 2019)—tangible entities with haptic interfaces and affordances that cater to possibilities for play, making them suitable tools for *edutainment* (entertainment with educational content) and playful learning. Commercial-level toy robots differ from other types of robots, such as assistant and service robots, in terms of their physicality, aesthetics, and functionality. They are relatively small, portable, and have a playful, often cute, and anthropomorphizable appearance that facilitates social interaction between the player and the robot (see Chapter Seventeen). This makes them *sociable machines* (Mascheroni

& Holloway, 2019) and relatable others (Li et al., 2020). Importantly, they are mainly designed to be playable and so channel invitations to play through sound, lights, and movement.

Often, smart toy robots link to the Internet: They are Internet-connected toys (IoToys) that combine different capabilities such as networking, processing, and interaction possibilities (see e.g., Sylla et al., 2022). Following a definition by Chaudron et al. (2017, p. 27), smart toys "contain embedded electronic features such that it can adapt to the action of the user," whereas connected toys "connect to a remote server that collects data and empowers the toys' intelligence." While smart toys do not necessarily enable a Wi-Fi connection, for IoToys their ability to connect online is an essential feature. The educational aspect is often accentuated in the marketing of smart toys, including toy robots (Ihamäki & Heljakka, 2018).

One example of an IoToy is the *Dash* toy robot from Wonder Workshop (for players of 6 years and up) that comes with a shiny plastic shell and is operated with a smartphone app. To steer Dash, the screen of the smartphone works as a remote control on which the coding exercises are performed. Another example of an Internet-connected toy robot is the Fisher-Price Smart Bear (for players aged 3–8 years): An "interactive learning friend with all the brains of a computer, without the screen," (Amazon, 2023) that has a soft, textile surface familiar with plush toys. For this toy robot, a free app provides unlimited Wi-Fi content that enables the "toy friend" (Heljakka & Ihamäki, 2019, p. 160) to communicate stories for players to listen to and challenges for them to solve during playtime. However, instead of the app, players use smart cards by showing them to the Smart Bear, which prompts various invitations to play with the toy.

DEFINING "TOY" AND "PLAY" IN RELATION TO ROBOTS

While there is a prevailing concern that technology use may diminish play (Marsh & Bishop, 2014), the most obvious modern manifestations of play as a consumable experience include children's toys, computers, and video games (Sutton-Smith, 2017, pp. 233–234). Toys are the things imagined, innovated, designed, and produced for play. Today, technologically-augmented toys come in many forms. Smart toy robots as sociotechnical entities are gaining a creature-like quality (Ackermann, 2005). For the remainder of this chapter, I will focus on a specific kind of toy robot: *Character toys* (Heljakka, 2013), meaning toys with a face, technological features, the ability to communicate and interact with their players, and the possibility to connect to online environments. Recently, three-dimensional character toys integrated with digital and connected technologies have been referred to as smart toys, high-tech toys, interactive toys, and robotic toys (ter Stal, 2017), which may be categorically understood as *digital toys*. Digital toys provide an interactive environment for children to experience novel technology (Sung, 2018).

Here, the attention turns to smart toy robots, with a particular focus on Internet-connected toys, which interact with players either through the components of the associated mobile device screens or interact as the complete, physical thing that the toy is. While mobile devices attract focus to their screens, the robot toy's haptics refocus attention on the toy part of technology: Technologically-augmented toys are haptic and are therefore "touchable" technology (Palaiologou et al., 2021). IoToys as a category of smart toy robots, provide sensory-rich interaction mediated through their curved surfaces, light-up features, and playful sounds, which impact possible play patterns that manifest in association with their use.

Technologically-augmented toys, such as the smart toy robots of interest for this chapter, derive meaning when being played with. There are many definitions of play. This chapter relies on a definition given by Plowman and Stephen (2014, p. 19): "Play is non-compulsory, driven by intrinsic motivation and undertaken for its own sake, rather than as a means to an end. It may take infinite forms, but the key characteristics of play are fun, uncertainty, challenge, flexibility, and non-productivity." Depending on the context—for example, leisure or learning—play with technologically-enhanced toys may take different forms. What separates toy play from other types of play (for example, gameplay) is the open-endedness of the scenarios. Games have goals that create a structure for play; toys do not demand such structure but instead, create space for the play to take form in the players' imagination. With emerging technologies, however, smart toy robots have started to illustrate game-like features, and inspire game-like play patterns (e.g., Ihamäki & Heljakka, 2018).

What differentiates technology-augmented toys from traditional (i.e., non-technological) toys is the interaction they provide: They react to the player's actions. Often, this interaction is mediated through a screen, operated through the use of a mobile device. Today, technology-augmented play often manifests through smartphones, tablets, computers, and consoles. Currently, we are in a situation where screen-focused devices, such as tablets, are already considered toys in their own right (Ihamäki & Heljakka, 2021). Yelland (1999) notes how the concept of "toy" has changed considerably as technologies have brought additional dimensions to objects that had previously been relegated to a passive role in their interactions with children (p. 217). For instance, due to technological-augmentation, toys like dolls and building blocks can communicate through language, lights, and movements.

AFFORDANCES AND PLAY EXPERIENCES

Toys (or, more precisely, their designers) seek to control play through *affordances* (Norman, 2007)—the cues through which a device communicates. "With products that require less knowledge for their operation, learning usually occurs by following cues within the interface" (Margolin, 2002). Following Norman, Wang et al. (2022) argue that the affordances of digital toys are of two types: Natural and deliberate. While natural affordances refer to inherent properties of objects (size, shape, or material), deliberate affordances are information incorporated into designs to direct people to discover possible actions. Interactive toys can produce new affordances compared to non-technological (i.e., traditional) toys (Thrift, 2003). Again, a toy with more affordances might have more play value because it could allow for more types of play (Kudrowitz & Follett, 2014).

The relevance of various toy affordances unfolds in the experiences gained during play. We may classify toy play as manifesting physical, functional, fictional, and affective experiences (Heljakka, 2018). The physicality (e.g., size, shape, thematic design) of a toy matters: For example, the use of different textures in toys points to the possibility of sensory-rich experiences. Functionality refers to the mechanical and increasingly digital operation of toys, which cater to possibilities for engaging with the toy in spatial and cognitive ways—that is, enabling movements of the body and mind. Further, toys provide entertainment and possibilities to learn: Many current toys include deliberately designed educational components such that they are sometimes understood as a form of education or *edutainment*. Finally, character toys usually have a fictional backstory, referring to a possible personality, which increases their potential for affective experiences by creating opportunities to bond with the toy as a character.

In toy robots, feedback and control are key features of interactive play (Kafai, 2021). In reference to gameplay, Flynn et al. (2019) divide the player-game interaction into four levels of engagement: *Receptive* interactivity (watching or listening to information given), *manipulative* interactivity (e.g., manipulating on-screen objects), *embodied* interactivity (using bodily motion to engage), and *contingent* interactivity, or exchanging a meaningful dialogue with an intelligent agent. All of these four modes of interactivity are relevant for interaction with smart toy robots as well: Players use their hands and bodies to interact with the smart toy robots and exchange meaningful dialogues with them, which provide content for play through sound, light, and movement.

IOTOY ROBOTS INTEGRATE TECHNOLOGY AND TOY

IoToy robots are a kind of smart toy robots, moving beyond the mediated, screen-based approaches described above to instead manifest a whole, embodied technology with Internet-connected operation. With that embodiment, IoToy robots have a three-dimensional presence and persuade players to interact through various affordances related to receptive, manipulative, embodied and contingent interaction. Like toys in general, IoToys may provide physical, functional, fictional, and affective play experiences. Physical experiences are delivered through their materiality (appearance, weight, and durability), functional experiences through mechanics (touch-sensitive surfaces), fictional experiences through storytelling capacities (producing content that augments its apparent personality), and affective experiences through emotional connection potentials (rounded shapes, plump bodies, welcoming posture, pleasant surface textures).

In the formulation of Fernaeus et al. (2010), robotic toys are intended for basic leisure activities such as play, creativity, playful learning, entertainment, and relaxation, and IoToys are no different. Active play is invited predominantly through physical and affective affordances of the IoToys (Berriman & Mascheroni, 2019). These *persuasive toy friends* (Heljakka & Ihamäki, 2019) may provide many forms of interaction attained through spatial and mobile engagement and possibilities for imaginative play. In our multi-stage research, we play-tested a set of IoToys with preschool children aged 5–6 years (Heljakka & Ihamäki, 2019; Ihamäki & Heljakka, 2021). The findings of the studies revealed how the affordances of IoToys employed in the study (the *Dash* robot from WonderWorkshop, Fisher-Price *Junior Smart Bear*, and CogniToys *Dino*) can be categorized into physical affordances (material dimensions and properties), pre-programmed (technological) affordances (sound, light, and movement, interactivity, and connectivity), which may be conceptualized as educational affordances (through the content the toys communicate, or invitations to play with stories, mini-games like coding exercises, tasks related to learning about good manners, etc.), and affordances for pretend play and personal meaning-making. The IoToys under scrutiny urged the children to *mobilize their imagination* and *advance their cognitive skills* through playing with sounds and language. Further, the exemplar robotic IoToys physically mobilized players by inviting them to *bodily movement*. Our observations follow Berriman and Mascheroni's assertion (2019) that future play experiences should consider gross motor activities such as running, throwing and catching to encourage active physical play over passivity fostered by other types of toys.

IOTOYS MAY BE FUTURE FRIENDS WITH AFFECTIVE RELATIONS

Robots are becoming more vital to society, which has consequences for the play lives of children (Lionel et al., 2020). Smart toy robots, including IoToys, offer multiple challenges, possibilities, and new directions for researching, designing, developing, deploying, regulating, engaging, and otherwise orienting ourselves in relation to robots. In parallel to providing possibilities for cognitive development and physical movement through interaction in play and playful learning, Internet-connected toys may become more like "relatable others" for children in the future. In fact, Peter (2017) discusses this as an ongoing *robotification of childhood*.

However, social robots may be entities that "can autonomously interact with humans in a socially meaningful way" (Chaudron et al., 2017, p. 7) and have tended to emulate affective relationships with humans—both children, and adults. According to Belpaeme et al. (2013), interactions between robots and children differ from those between adults and robots. Children treat robots as more than mere artifacts (Melson et al., 2009) such that presenting robots as autonomous agents may help promote children's social-emotional development (Chernyak & Gary, 2016). The tendency to attribute human-like qualities to robots, and train pro-social skills such as empathy with them is highly relevant when considering play and smart-toy friends from a broader perspective on players than children alone. We are invited and persuaded to become friends with these machines. Consequently, "Robots have become our indispensable mechanical companions" (Pesce, 2000, p. 220). To exemplify, we have observed IoToys, moving players emotionally by persuading them to form relationships with and nurture them (Heljakka & Ihamäki, 2019). Our longitudinal study revealed how "[T]he affective experience with Fisher-Price Smart Toy Bear enabled pleasurable experiences for the preschoolers when they nurtured and played house with the toy. Children received affective experiences when Dash made eye contact with them" (Ihamäki & Heljakka, 2021, p. 200). The tendency to establish relationships with smart toy robots is expected to become more significant in future development as envisioned in speculative toy fiction (Heljakka, 2022, 2023).

Speculative media texts represent the *technological imaginary* derived from popular storytelling (Giddings, 2019) and reveal how social robots of the future may look and be experienced. One example of toy-like robots featured in cinematic films include Disney-Pixar's *WALL-E*—infused by his creators with anthropomorphic features: Expressive eyes, performed fear, and a demonstrated yearning for connection (Tranter & Sharpe, 2011). Another is the connected toy robot character "Ron" in the film *Ron's Gone Wrong* (Smith et al., 2021). Ron is a "B*bot"—a speculated IoToy of the future that interacts with humans through a

360-degree projection screen (Heljakka, 2022; 2023). This means that the complete body of the envisioned toy robots consists of a screen, which challenges the notion of a flat-screen familiar from mobile device use and gameplay. "The B*bot has an infinite number of downloadable skins, apps, contacts, photos, chats, music, too" (Phegley, 2021, p. 1), illustrating affordances and content for play accessed through the body of the robot, which may be operated and interacted with through voice control. Further, the fictitious B*Bots also connect to social media platforms, mediating socially shared content through their projective bodies. In this way, the components of screen and links converge in the speculated toy robot, which gives rise to enhanced possibilities for physical movement and emotional bonding experienced in relation to developed social robots. Beyond these physical, functional, and fictional dimensions, B*Bots promise to be "a new kind of friend for everyone," which opens up the possibility for deeper and more meaningful affective relationships to build up between human players and their smart toy robots.

Krueger and colleagues (2021) claim that "Robotic agents will be life-long companions of humans in the foreseeable future. To achieve such successful relationships, people will likely attribute emotions and personality, assign social competencies, and develop a long-lasting attachment to robots" (p. 371). Embedded in the screens and network-linking connectivity of future toy robots, then, is an important question regarding the potential for enduring relationships with robots

Figure 10.1: "Ron" the B*bot, a speculative IoToy that interacts with humans through a 360-degree projection screen (source: Katriina Heljakka).

and consequently the possibility of mitigating consequences of human loneliness, a major concern in contemporary society: "Can toy robots become more than play partners—our social companions or even friends?"

REFERENCES

Ackermann, E. (2005). Playthings that do things; A young kid's "incredibles"! In *Proceedings of the Conference on Interaction Design and Children* (pp. 1–8). ACM.

Amazon. (2023). Fisher-Price Smart Bear. <https://www.amazon.com/Fisher-Price-DNV31-Smart-Bear/dp/B013UIQB9W>

Belpaeme, T., Baxter, P., De Greeff, J., Kennedy, J., Read, R., Looije, R., Neerincx, M., Baroni, I., & Zelati, M. C. (2013). Child-robot interaction: Perspectives and challenges. In *Proceedings of the International Conference on Social Robotics* (pp. 452–459). Springer.

Bergen, D. (2015). Psychological approaches to the study of play. *American Journal of Play, 7*(2), 51–69.

Berriman, L., & Mascheroni, G. (2019). Exploring the affordances of smart toys and connected play in practice. *New Media & Society, 21*(4), 797–814.

Chernyak, N., & Gary, H. E. (2016). Children's cognitive and behavioral reactions to an autonomous versus controlled social robot dog. *Early Education and Development, 27*(8), 1175–1189.

Chaudron, S., Di Gioia, R., Gemo, M., Holloway D., Marsh, J., Mascheroni, G., Peter, J., & Yamada-Rice, D. (2017). *Kaleidoscope on the Internet of Toys: Safety, security, privacy and societal insights*. Publications Office of the European Union.

Fernaeus, Y., Håkansson, M., Jacobsson, M., & Ljungblad, S. (2010). How do you play with a robotic toy animal? A long-term study of Pleo. *Proceedings of the International Conference on Interaction Design and Children* (pp. 39–48). ACM.

Flynn, R. M., Richert, R., A., & Wartella, E. (2019). Play in a digital world: How interactive digital games shape the lives of children. *American Journal of Play, 12*(1), 54–73.

Giddings, S. (2019). Toying with the singularity: AI, automata and imagination in play with robots and virtual pets. In G. Mascheroni & D. Holloway (Eds.), *The Internet of Toys* (pp. 67–87). Palgrave Macmillan.

Druga, S., Vu, S. T., Likhith, E., & Qiu, T. (2019). Inclusive AI literacy for kids around the world. In *Proceedings of FabLearn* (pp. 104–111). ACM.

Heljakka, K. (2013). *Principles of adult play(fulness) in contemporary toy cultures—From wow to flow to glow* [Doctoral thesis, Aalto University].

Heljakka, K. (2018). Dimensions of the toy experience. In J. Paavilainen, K. Heljakka, J. Arjoranta, V. Kankainen, L. Lahdenperä, E. Koskinen, … & R. Koskimaa. (Eds.), *Hybrid social play final report*. University of Tampere.

Heljakka, K. I. (2022, November). Reading Ron right: Speculative toy fiction, friendship and design of future IoToys. In *Proceedings of the International Academic Mindtrek Conference* (pp. 334–338). ACM.

Heljakka, K. (2023). Moved by B*Bots: Speculative toy fiction and play with future IoToys. In *Proceedings of the Future Technologies Conference* (pp. 154–173). Springer.

Heljakka, K., & Ihamäki, P. (2019). Persuasive toy friends and preschoolers: Playtesting IoToys. In G. Mascheroni & D. Holloway (Eds.), *The Internet of Toys: Practices, affordances and the political economy of children's play* (pp. 159–178). Palgrave Macmillan.

Ihamäki, P., & Heljakka, K. (2018). Smart toys for game-based and toy-based learning—A study of toy marketers', preschool teachers' and parents' perspectives on play. *Proceedings of the International Conference on Advances in Human-oriented and Personalized Mechanisms, Technologies, and Services* (pp. 14–18). IARIA.

Ihamäki, P., & Heljakka, K. (2021). Internet of Toys and forms of play in early education: A longitudinal study of preschoolers' toy-based learning experiences. In D. Holloway, M. Willson, K. Murcia, C. Archer, & F. Stocco (Eds.), *Young children's rights in a digital world: Play, design and practice* (pp. 193–204). Springer.

Kafai, Y. B. (2021). Play and technology: Revised realities and potential perspectives. In *Play from birth to twelve and beyond* (pp. 93–99). Routledge.

Krueger, F., Mitchell, K., C., & Gopikrishna, D., & Katz, J. S. (2021). Human–dog relationships as a working framework for exploring human–robot attachment: A multidisciplinary review. *Animal Cognition, 24*, 371–385.

Kudrowitz, B., & Follett, J. (2014). Emerging technology and toy design. In *Designing for Emerging Technologies* (pp. 237–254). O'Reilly.

Li, Z., Cummings, C., & Sreenath, K. (2020, October). Animated Cassie: A dynamic relatable robotic character. In *Proceedings of the International Conference on Intelligent Robots and Systems* (pp. 3739–3746). IEEE.

Lionel P. R. Jr., Rasha, A., Esterwood, C., Sangmi, K., Sangseok, Y., & Zhang, Q. (2020). A review of personality in human–interactions. *Foundations and Trends in Information Systems, 4*(2), 107–212.

Margolin, V. (2002). *The politics of the artificial: Essays on design and design studies*. The University of Chicago Press.

Marsh, J., & Bishop, J. C. (2014). *Changing play: Play, media and commercial culture from the 1950s to the present day*. Open University Press/McGrawHill.

Mascheroni, G., & Holloway, D. (2019). Introducing the internet of toys. In *The internet of toys: Practices, affordances and the political economy of children's smart play* (pp. 1–22). Palgrave.

Melson, G. F., Kahn, P. H. Jr., Beck, A., & Friedman, B. (2009). Robotic pets in human lives: Implications for the human–animal bond and for human relationships with personified technologies. *Journal of Social Issues 65*(3), 545–567.

Norman, D. (2007). *Emotional design: Why we love (or hate) everyday things*. Basic Books.

Palaiologou, I., Kewalramani, S., & Dardanou, M. (2021). Make-believe play with the Internet of Toys: A case for multimodal playscapes. *British Journal of Educational Technology, 52*(6), 2100–2117.

Pesce, M. (2000). *The playful world: How technology is transforming our imagination*. Ballantine.

Peter, J. (2017). Social robots and the robotification of childhood. In Chaudron S., Di Gioia, R., Gemo, M., Holloway, D., Marsh, J., Mascheroni, G., Jochen, P., & Yamada-Rice, D. (Eds.), *Kaleidoscope on the Internet of Toys—Safety, security, privacy and societal insight.* (pp. 14–16). Publications Office of the European Union.

Phegley, K. (2021). *Ron's gone wrong: The official movie novelization*. Penguin Young Readers Licenses.

Plowman, L., & Stephen, C. (2014). Digital play. *The SAGE handbook of play and learning in early childhood* (pp. 330–341). SAGE.

Smith, S., Vine, J.-P., & Rodriguez, O. E. (Directors). (2021). *Ron's gone wrong* [Film]. 20th Century Studios.

Sung, J. (2018). How young children and their mothers experience two different types of toys: A traditional stuffed toy versus an animated digital toy. *Child Youth Care Forum, 47*, 233–257.

Sutton-Smith, B. (2017). *Play for life: Play theory and play as emotional survival.* The Strong.

Sylla, C., Heljakka, K., Catala, A., & Ozgur, A. G. (2022). Smart toys, smart tangibles, robots and other smart things for children. *International Journal of Child-Computer Interaction, 33*, 100489.

ter Stal, S. (2017). Designing and interactive storytelling system for children using a smart toy. Design of a prototype to investigate the effect of emotional behaviour of a toy on children's storytelling [Master's thesis, University of Twente].

Thrift, N. (2003). Closer to the machine? Intelligent environments, new forms of possession and the rise of the supertoy. *Cultural Geographies 10*, 389–407.

Tranter, P., & Sharpe, S. (2011). Disney-Pixar to the rescue: Harnessing positive affect for enhancing children's active mobility. *The Journal of Transport Geography 20*, 34–40.

Wang, Y., Vickery, N. E., Tarlinton, D., Ploderer, B., Knight, L., Blackler, A., & Wyeth, P. (2022). Exploring the affordances of digital toys for young children's active play. In *Proceedings of the Australian Conference on Human-Computer Interaction* (pp. 325–337). ACM.

Yelland, N. (1999). Technology as play. *Early Childhood Education Journal, 26*, 217-220.

CHAPTER ELEVEN

Memory & Information: The Core of Robot Cognition

RAFAEL SOUSA SILVA & TOM WILLIAMS[1]

Robots and humans must deal with large amounts of information in order to navigate the complexities of daily activities, including the processing of visual, aural, haptic, and kinesthetic data. To better understand how robots engage that information, it can be useful to draw on knowledge of how the human cognitive system processes and acts on information—that is, the way human knowledge is obtained, communicated, and understood (Buckland, 1991). This information is encoded, stored, and retrieved through memory. Memory plays a critical role in many important human cognitive processes, allowing the recall of previous events and the formation of links between previously obtained information and ongoing or subsequent stimuli. Given its key role in human cognition (see Chapters 22–23), it is common to find psychologically-inspired models of memory used in the field of cognitive robotics. In fact, in recent years memory has started to transition from a passive background storage mechanism to a core process that guides robot cognitive systems. As such, researchers are beginning to explore robot cognitive systems through memory-centered approaches in which memory is considered to be the substrate of cognition and to have a pervasive role in guiding different aspects of robot behavior (Baxter & Browne, 2010; Baxter et al., 2011; Baxter & Belpaeme, 2014). Memory has a great impact, for example, on multiple processes related to human natural language, such as language understanding, acquisition,

[1] *This work was funded in part by NSF CAREER grant IIS-2044865.*

and referring expression generation (Rönnberg et al., 2010; Baddeley et al., 1998; Gundel et al., 1993). These processes are also very important for social robots that will be deployed in interactive real-world environments.

One of the most common ways to taxonomize both human and computational models of memory is to subdivide them into long-term memory and short-term memory according to overall storage capacity and the duration for which information remains available (Hebb, 2005). On one hand, long-term memory can keep items in storage for extended amounts of time, and it is the memory subsystem responsible for handling durable knowledge from previously experienced events as well as knowledge about how to perform skills and abilities. On the other hand, short-term memory has a limited capacity and information within it is subject to a more immediate process of forgetting. Both long-term memory and short-term memory are critical for robots intended to engage in social interactions with human users. In human-robot interaction scenarios, for instance, robots may need explicit long-term memory systems to remember information from previous encounters with humans in order to carry out interactions and conversations that fit with and respect that interaction history, and that prevent interactants from needing to repeat themselves (Jokinen & Wilcock, 2021). For instance, a math robot tutor may remember which problems were covered with a specific student to avoid redundant sessions. Similarly, robots may need explicit short-term memory systems to maintain awareness of the most critical elements of an unfolding interaction context, and to leverage that awareness to align with human expectations. A robot tutor may, for example, facilitate interaction by keeping in short-term memory the salient details of a problem that is currently being tackled. A well-designed memory system should thus be understood as one of the most critical mechanisms in the cognitive architectures of social robots. In the next sections, we will dive deeper into some of the specific design choices important to robot memory systems.

LONG-TERM MEMORY

Long-term memory is responsible for the management of information that is retained for extended or undetermined amounts of time. In the field of robotics, long-term memory often works through a set of knowledge bases (i.e., databases where information and rules about the world are stored). A robot will access this information and use it as reference for subsequent tasks and actions that need to be performed. Although memory has been a hot topic of discussion and research in cognitive psychology for over a century, it is still at an early stage in the field of robotics. Consequently, many aspects of robotic long-term memory are only now beginning to draw more inspiration from models of human memory to promote better social interactions. In this section we will focus on three important aspects of long-term memory for robots that are being tackled by current research and will lead the way to future work on memory for artificial social agents—knowledge domains, information distribution, and information representation.

Which Domains of Knowledge Are Important for Social Robots?

Social robots need to organize their knowledge according to the relevant features of their tasks and of the social contexts in which they participate. Therefore, it is important for robotic systems to avoid simply storing all the information they can, as this approach will likely clutter knowledge bases with irrelevant information that will never be used, thus slowing down robots' cognitive processes. But how may roboticists determine which kinds of information are the most relevant for social robots? One possible answer may be to borrow inspiration from human models of cognition. For example, a promising idea can be found in the Theory of Core Knowledge proposed by Spelke and Kinzler (2007). This theory proposes that, at its early stage, human knowledge is distributed across five base domains: People, locations, numbers, actions, and objects. These domains then serve as foundation for cognitive processes, such as language acquisition and the development of social skills. Through the same lens, it is possible to imagine social robots that split long-term information into these core domains and use this knowledge to inform key processes, such as those related to language or reasoning. Borrowing inspiration from human models of cognition such as the Core Knowledge model is an important approach for the implementation of social robots, as it may promote more natural, relevant, and intuitive interactions.

How Can Long-term Information Be Organized?

Another important question regards the different ways in which information can be distributed across a robot's architecture. Because of the common implementation of memory as a passive storage mechanism, many early cognitive models tended to condense an agent's acquired knowledge into a single or a small number of knowledge bases. This reduces the amount of information endpoints an artificial agent needs to consult when specific information is queried. However, this also creates a bottleneck for information access and limits the robotic agent's capacity to process information. To resolve this issue, some cognitive systems split long-term memory across multiple, distributed knowledge bases according to the type of objects or events that information relates to (cf., Williams & Scheutz, 2016). The presence of multiple knowledge bases creates more access channels when information needs to be retrieved, allowing for better architectural efficiency through reduced algorithmic runtimes. It also allows for different types of information to be represented in the ways that suit each domain best, which may depend, for instance, on the data structures chosen to represent and store each type of information. Our example robot tutor, for instance, may represent procedural information through a commonly used planning language (e.g., Planning Domain Definition Language, or PDDL), may represent information related to objects and people as images, and may represent information about locations through a graph and/or occupancy grid. This leads us to our last topic of discussion, which is centered around these different types of information representation.

What Is the Best Way to Represent Long-term Information?

Recent research has proposed that, in addition to borrowing inspiration from memory taxonomy, synthetic long-term memory models should take into account the context in which they are constructed (Wood et al., 2012). Thus, to determine the best way to represent long-term information, it is important to understand the tasks a social robot will perform and the interactive settings in which it will participate. Long-term memory can be subdivided into procedural memory and declarative memory, and the latter can be further split into episodic memory and semantic memory. While each subsystem is tied to processes that are important to an agent's cognition, episodic memory is specifically important to social robotics, as it allows agents to remember the spatiotemporal details (i.e., the qualities of time and space) composing the context of specific events from the past. They can then use this prior knowledge to inform new interactions. For example, episodic information may tell our math robot tutor that it has already covered Chapter Three questions with the current student last week, so they can resume study from Chapter Four. Given the importance of episodic memory, in recent years more emphasis has been placed on the different ways that episodes can be represented to facilitate recall (Rothfuss et al., 2018). The most common way of representing knowledge in computational models of long-term memory is through symbolic representations, such as first-order logic statements (e.g., *green*(X) to represent that something is green, or *in*(X, Y) to represent that object X is inside object Y). This kind of representation allows the easy formulation of definitions for different types of events and objects, and it allows the derivation of rules based on statements stored within knowledge bases (Wood et al., 2012). However, if we borrow inspiration from human models of cognition, we may find ways to complement these symbolic representations and create better snapshots of experienced events. For instance, our robot tutor could build more robust study-session snapshots by storing the page number associated with a problem set, pictures of the student's reactions to a problem explanation, and the logical representations of the text associated with each question. There are multiple different ways in which long-term information and episodes can be represented, yet the way a social robot represents information should be rooted in the interactive scenarios in which it will participate.

SHORT-TERM MEMORY

While answering these initial questions about long-term memory will lead to insightful advancements in social robotics, it is also important to understand how long-term memory may work in equilibrium with short-term memory. In robotic cognitive systems, short-term memory models can influence what happens within long-term memory itself and may contribute to more efficient robot architectures. In cognitive psychology, the concept of short-term memory has evolved from a unitary-system perspective to a multi-modal system known as working memory. The most widely accepted model of working memory was proposed by Baddeley and Hitch (1974). Their model subdivided short-term memory into a central component and three subsystems: The visuospatial sketchpad (that supports mental models for spaces), the phonological loop (processing verbal information; see Chapter Nine), and the episodic buffer (integrating long- and short-term memory information). Each of these subsystems is responsible for the management and storage of different types of recently acquired information. As such, working memory is the multi-component memory subsystem responsible for managing information obtained from immediate or ongoing cognitive tasks. It is a central structure that informs essential cognitive processes, such as reasoning (Kyllonen & Christal, 1990), comprehension (Halford et al., 1998), and learning (Baddeley, 2010). Given its importance to human cognition, working memory has been a key topic of discussion within the cognitive psychology community for many decades and diverging theories have been proposed. Yet, across these diverse models of working memory, a few characteristics seem to be widely accepted by researchers. First, it has limited capacity, with early research presenting estimations in the range from four to nine items (Cowan, 2001; Miller, 1956). Second, the information within it is volatile and may be forgotten at faster rates than information contained inside of long-term memory (Waugh & Norman, 1965). Finally, working memory contents are readily accessible to other cognitive processes and may immediately influence deliberative processes (Norman & Shallice, 1986).

In robotics, some researchers have used this set of widely accepted working memory features to guide their computational models, aiming to achieve better system efficiency, implement human-like behavior, and replicate core features of human cognition. Architectural implementations of working memory are often directed at improving robots' decision-making processes, helping with the selection of action plans that are appropriate to the situation in which a robotic agent is immersed. Work done with the Adaptive Control of Thought–Rational (ACT-R) and Soar cognitive architectures (Anderson et al., 1996; Nuxoll et al., 2004), for example, is based on theories of activation from cognitive psychology, in which working memory is represented as the set of context-relevant, salient (i.e., activated) information at any given time. Although these models reflect cognitive psychology theories based on *temporal decay* (i.e., information is forgotten over time when not rehearsed; Ebbinghaus, 2013; Brown, 1958), other architectures implement complementary models to guide working memory. For example, recent research on the Distributed, Integrated, Affect, Reflection, and Cognition (DIARC) architecture (Sousa Silva et al., 2023) implements a psychologically inspired model of *retroactive interference* in which the similarity between two pieces of information can interfere with recall and older items are replaced with newer ones (Waugh & Norman, 1965; Dewar et al., 2007). This interference dynamic directly informs utterances that are generated by social robots during conversation. In addition to these implementations based on the forgetting dynamics that define working memory, current models also treat working memory as a goal-driven working space (Wood et al., 2012). This approach treats working memory as a hub for the most relevant information given the agent's current task and focus of attention. In essence, current models of robotic working memory tend to borrow inspiration from the way human working memory functions. As the widely accepted working memory dynamics and the goal-driven approaches continue to guide new research, robotic models continue to advance towards more efficient and human-like memory-centered behavior.

WHAT COMES NEXT FOR MEMORY?

As we transition away from models in which memory acts as a simple, unitary "information buffer" and toward a well-structured, multi-component system capable of handling procedural, semantic, and episodic information appropriately (Wood et al., 2012; Baddeley & Hitch, 1974), we may assess the extent to which certain constraints that define human memory need to be respected. While the implementation of memory models directly inspired by human memory seems promising, it may not be the gold standard for promoting more natural and human-like interactions with robots. Perhaps leveraging the capabilities and advantages of computational systems by relaxing some of the constraints on human memory can lead to the development of more efficient and more useful social agents. For instance, should robotic working memory have a maximum storage capacity for activated features? If so, should the upper bound be the same as that for humans or can it be higher? With the advancement of cloud systems, in which the amount of information that can be stored is practically unlimited, we might find a solution to the problem of creating systems that are restricted to a certain amount of information. This idea of "superhuman" or exceptional memory, which has become a stereotypical characteristic that defines robots from movies and general media (see Chapter Fifteen), may lead to interesting advances in robotics, such as allowing our example robot tutor to remember all past student reactions to a specific problem (as opposed to only a subset of such reactions) and using that knowledge to inform a new problem explanation.

However, that exceptional nature could change the way these agents are socially perceived, dividing public opinion about how and where these agents should operate. It is important to consider how advances in memory research might affect the many dimensions of human-robot interaction scenarios (e.g., trust, agency, power), if at all. Are humans going to be comfortable with robots who remember information for extended periods of time? How should we determine what a social robot is and is not allowed to remember? Do we want robots to remember and store important and sensitive information, such as people's faces, addresses, relationships, and habits? Considering these questions is crucial to avoid entering the typical dystopian portrayals of technologically advanced societies on the edge of collapse that we are used to seeing on media channels. Most importantly, and independently of whether models should be directly inspired on human cognition, these robotic systems will need to address the important privacy and ethical concerns that may arise from advancements in the field. Specifically, they will need not only to be transparent, but also to account for important technological literacy and equitable design processes (cf. European Commission, 2019; Hemment et al., 2023; Long & Magerko, 2020; Mott & Williams, 2023; Wortham et al., 2016). Fulfilling these requirements is important, as they inform those who will regularly interact with these social agents about how their information is being handled and how the robotic system was implemented in the first place. In addition, it is important for mitigating problematic biases that may be carried into the development of such systems and for promoting more inclusive human-robot interaction scenarios.

In conclusion, memory is central to humans and has a direct impact on many important processes of cognition. Consequently, it should also be a central aspect of cognition within robotic systems. The careful implementation of memory models is key and will provide invaluable insights to the next generations of social agents in robotics. However, advancements in memory-centered systems will lead to important questions about privacy and ethical implications that may arise from their gradual introduction to society. The answers to these questions should aim to preserve and respect humanity and can lead the way to social agents who will add to society and be better prepared to handle open-world interactions with humans. As memory systems evolve, we should thus think carefully about the social implications they will introduce in order to promote a responsible usage of artificial social agents.

REFERENCES

Anderson, J., Reder, L., & Lebiere, C. (1996). Working memory: Activation limitations on retrieval. *Cognitive Psychology, 30*(3), 221–256.

Baddeley, A., & Hitch, G. (1974). Working memory. *Psychology of Learning and Motivation, 8*, 47–89.

Baddeley, A., Gathercole, S., & Papagno, C. (1998). The phonological loop as a language learning device. *Psychological Review, 105*(1), 158.

Baddeley, A. (2010). Working memory. *Current Biology, 20*(4), R136–R140.

Baxter, P., & Belpaeme, T. (2014). Pervasive memory: The future of long-term social HRI lies in the past. In *Proceedings of the International Symposium on New Frontiers in Human-Robot Interaction* (pp. 1–4). AISB.

Baxter, P., & Browne, W. (2010). Memory as the substrate of cognition: A developmental cognitive robotics perspective. In *Proceedings of the International Conference on Epigenetic Robotics* (pp. 19–26). Örenäs Slott.

Baxter, P., Wood, R., Morse, A., & Belpaeme, T. (2011, November). Memory-centred architectures: Perspectives on human-level cognitive competencies. In *Proceedings of the Association for the Advancement of Artificial Intelligence Fall Symposium* (pp. 26–33). AAAI.

Brown, J. (1958). Some tests of the decay theory of immediate memory. *Quarterly Journal of Experimental Psychology, 10*(1), 12–21.

Buckland, M. (1991). Information as thing. *Journal of the American Society for Information Science, 42*(5), 351–360.

Cowan, N. (2001). The magical number 4 in short-term memory: A reconsideration of mental storage capacity. *Behavioral and Brain Sciences, 24*(1), 87–114.

Dewar, M., Cowan, N., & Della Sala, S. (2007). Forgetting due to retroactive interference: A fusion of Müller and Pilzecker's (1900) early insights into everyday forgetting and recent research on anterograde amnesia. *Cortex, 43*(5), 616–634.

Ebbinghaus, H. (2013). Memory: A contribution to experimental psychology. *Annals of Neurosciences, 20*(4), 155.

European Commission. (2019). Ethics guidelines for trustworthy AI [report]. <https://digital-strategy.ec.europa.eu/en/library/ethics-guidelines-trustworthy-ai>

Gundel, J., Hedberg, N., & Zacharski, R. (1993). Cognitive status and the form of referring expressions in discourse. *Language, 69*(2), 274–307.

Halford, G., Wilson, W., & Phillips, S. (1998). Processing capacity defined by relational complexity: Implications for comparative, developmental, and cognitive psychology. *Behavioral and Brain Sciences, 21*(6), 803–831.

Hebb, D. (2005). *The organization of behavior: A neuropsychological theory*. Psychology Press.

Hemment, D., Currie, M., Bennett, S., Elwes, J., Ridler, A., Sinders, C., ... & Warner, H. (2023, June). AI in the public eye: Investigating public AI literacy through AI art. In *Proceedings of the Conference on Fairness, Accountability, and Transparency* (pp. 931–942). ACM.

Jokinen, K., & Wilcock, G. (2021, August). Do you remember me? Ethical issues in long-term social robot interactions. In *Proceedings of the International Conference on Robot & Human Interactive Communication* (pp. 678–683). IEEE.

Kyllonen, P., & Christal, R. (1990). Reasoning ability is (little more than) working-memory capacity?! *Intelligence, 14*(4), 389–433.

Long, D., & Magerko, B. (2020, April). What is AI literacy? Competencies and design considerations. In *Proceedings of the Conference on Human Factors in Computing Systems* (pp. 1–16). ACM.

Miller, G. (1956). The magical number seven, plus or minus two: Some limits on our capacity for processing information. *Psychological Review, 63*(2), 81–97.

Mott, T., & Williams, T. (2023). Rube-Goldberg machines, transparent technology, and the morally competent robot. In *Companion of the International Conference on Human-Robot Interaction* (pp. 634–638). ACM/IEEE.

Norman, D., & Shallice, T. (1986). Attention to action: Willed and automatic control of behavior. In *Consciousness and self-regulation: Advances in research and theory*, Vol. 4 (pp. 1–18). Boston, MA: Springer US.

Nuxoll, A., Laird, J., & James, M. (2004, September). Comprehensive working memory activation in Soar. In *Proceedings of the International Conference on Cognitive Modeling* (pp. 226–230). Psychology Press.

Rönnberg, J., Rudner, M., Lunner, T., & Zekveld, A. (2010). When cognition kicks in: Working memory and speech understanding in noise. *Noise and Health, 12*(49), 263–269.

Rothfuss, J., Ferreira, F., Aksoy, E., Zhou, Y., & Asfour, T. (2018). Deep episodic memory: Encoding, recalling, and predicting episodic experiences for robot action execution. *IEEE Robotics and Automation Letters, 3*(4), 4007–4014.

Sousa Silva, R., Lieng, M., & Williams, T. (2023). Forget about it: Entity-level working memory models for referring expression generation in robot cognitive architectures. In *Proceedings of the Annual Meeting of the Cognitive Science Society* (Vol. 45, No. 45). CSS.

Spelke, E., & Kinzler, K. (2007). Core knowledge. *Developmental Science, 10*(1), 89–96.

Waugh, N., & Norman, D. (1965). Primary memory. *Psychological Review, 72*(2), 89–104.

Williams, T., & Scheutz, M. (2016, March). A framework for resolving open-world referential expressions in distributed heterogeneous knowledge bases. In *Proceedings of the AAAI Conference on Artificial Intelligence* (Vol. 30, No. 1). AIII.

Wood, R., Baxter, P., & Belpaeme, T. (2012). A review of long-term memory in natural and synthetic systems. *Adaptive Behavior, 20*(2), 81–103.

Wortham, R., Theodorou, A., & Bryson, J. (2016). What does the robot think? Transparency as a fundamental design requirement for intelligent systems. In *Proceedings of the IJCAI Workshop on Ethics for Artificial Intelligence: Joint International Conference*. IJCAI.

CHAPTER TWELVE

Sensors & Actuators: Synergies of Perception and Movement

UCHENNA OGENYI

The robot companion Aibo—a charming robotic dog by Sony—has a range of embedded sensors (i.e., information collectors) that enable it to recognize faces, react to soft touches, and even understand spoken languages. It also has a system of actuators (i.e., movement generators) that help to puppeteer this electronic canine's movements, such as wagging its tail in delight, gracefully rolling over, or playfully twisting about. These complex behaviors are actually the result of *synergies* between the sensors and actuators as they support robots' abilities to interact seamlessly with their surroundings. These components and their coordination are the very essence of how robots function and operate. In a way, sensors serve as the robot's sensory organs enabling it to access its surroundings, while actuators are the motors behind the robot's joints that are responsible for translating control signals into physical motions. In their synergy, the information that actuators receive from sensors directs their actions.

SENSORS IN SOCIAL ROBOTIC SYSTEMS

Sensors are devices or components of the robot designed to capture, detect, or measure physical properties or information (e.g., color) related to a target (e.g., an object) and convert this data into readable or applicable outputs. Robots utilize sensors to receive or channel information from the surroundings and other objects they encounter. This data can be processed by the robot's control system (see Chapter Twenty Three) to allow the robot to make decisions and respond to changes even in real-time. Sensors have the capability to identify various stimuli such as light, sound, temperature, pressure, proximity, and others. This enables robots to autonomously navigate, avoid obstacles, manipulate objects, and execute a broad range of tasks.

Robots are designed for various purposes and the sensors incorporated into them from manufacturing stage are tailored to meet the requirements of those tasks. For example, a robot designed for industrial assembly might need proximity sensors for object detection and ultrasonic sensors for distance measurement, while a robot designed for exploring unknown environments might be equipped with cameras and environmental sensors (e.g., laser sensors) for effective navigation. Hence, the selection of sensors depends mainly on the specific needs and objectives of the robot's intended application. However, manufacturers sometimes provide the option to integrate additional sensors into the robot after production. This process often requires the assistance of an expert to ensure proper integration. For example, the disc-shaped TurtleBot robot (Clearpath Robotics, n.d.) comes with an open-source platform (meaning that anyone can modify the robot's design, software, and hardware specifications), and interface that allows additional sensors to be integrated into the robot in order to enhance its capabilities and to meet their needs.

Sensors are often categorized based on the type of physical quality (i.e., measurable property) they are designed to detect or measure, or their area of application. This classification helps in understanding the specific function and application of each type of sensor. For the purpose of social applications, I will categorize sensors based on the modalities of information they collect: Vision, audio, proximity, and tactile.

Vision Sensors

Vision-based sensors are the type of sensor that captures visual information (i.e., data or information acquired through the sense of sight) about their surroundings or objects of interest. This visual information generally refers to images or video streams captured by the vision sensor. Vision sensors play an important role as they enable robotic systems to see and understand their surroundings (Li et al., 2019). A common example of a vision sensor is the low-cost RGB camera (Tashtoush et al., 2021). RGB stands for red, green, and blue which are the primary colors of light (meaning that the three colors can be used to make other colors). The RGB camera operates by measuring the light intensity in each of the RGB channels and combining them to ascertain the precise color of the target object. Due to its ease of use and affordability, the RGB camera has been widely adopted for various robotic applications, including object detection (i.e., identification and localization of objects of interest within an environment) and tracking moving objects

(Nguyen et al., 2018). Furthermore, data captured by visual sensors are critical to social robots because it can be further processed by using some machine learning algorithms (i.e., computational techniques that enable a robot to learn from data and make decisions) to allow the robot to recognize human actions (Lee & Ahn, 2020). These actions encompass a wide spectrum, including but not limited to facial recognition, emotion detection, object recognition, gesture recognition, and visual tracking.

Audio Sensors

Audio sensors are sensors used for detecting sound waves in an environment. These sensors work by capturing sound waves that make the diaphragm (i.e., flexible membrane used in microphones) of the sensor vibrate through air pressure. Then through the transduction process (i.e., conversion of one form of energy into another form), the vibration is turned into an electrical signal which could be processed and analyzed by the robotic system. The audio information can be utilized in various robotic applications. One notable application is sound localization (Rascon & Meza, 2017). This allows a robot to determine the direction from which sound originates, providing a spatial awareness which fosters dynamic and responsive exchange for a robot and human voice interaction. This is because, based on the spatial awareness, a robot can interpret vocal cues of a human interactor, understand their intentions, and respond appropriately (Fiore, 2013). Because audio sensors are instrumental in robots' abilities to engage in natural-language exchanges, contributing to a more intuitive and user-friendly interaction experience (Marge et al., 2022).

Proximity Sensors

Proximity sensors are kind of sensors that detect the presence or absence of a material item or object within a physical range. Their detection range depends on the distance capacity, which is defined at the manufacturing stage. The main data that a proximity sensor gathers is the distance between itself and the detected object. Proximity sensors can be designed using different technologies, which determine their object-detecting principles. For example, a proximity sensor built using ultrasonic sensing technology operates by emitting sound waves and measuring the time it takes for the waves to bounce back after hitting an object. By knowing the speed of sound and the time it takes for the waves to return, the sensor calculates the distance. These sensors allow the robot to detect the presence of nearby objects which provides the robot with an ability to adapt to the environment (Moller et al., 2021). This sensor has found applications in social robotic systems. For instance, let's consider a robotic vacuum cleaner fitted with proximity sensors. These sensors enable obstacle detection which allows the robot to adapt its cleaning path such that, when it encounters a furniture, the robot intelligently calculates its distance and navigates around it, ensuring thorough cleaning without collisions.

Tactile Sensors

Tactile sensors are sensors that can detect and measure physical contact or touch. This is possible because they are designed to detect some physical parameters such as pressure, force, or vibration applied to their surfaces. When pressure is applied to the sensor, the material or mechanism within the sensor undergoes a change. This change alters the electrical properties of the sensor, causing it to generate a signal. The signal can vary depending on the type of sensing materials (e.g., conductive polymers, or resistive films) used to design the sensor. Tactile sensors are commonly used in robotic applications to provide feedback or control based on touch stimuli. An example of such application is in object gripping, where the tactile sensor is integrated into the gripper of the robot. This is then used to detect the force and pressure applied when gripping objects, enabling the robot to grasp objects securely without damaging them. It can also allow robots to adjust their grip strength according to the object's properties, perhaps delicately holding a lightweight sheet of paper but more firmly gripping a heavier sheet of metal. Another implementation of tactile sensors sees them embedded in robotic skin technology (Kappassov et al., 2015). The robotic skin can enable the measurement of further information such as texture and object deformation arising from surface touch of the object in contact with. This additional tactile information can be used to enable a robotic system to perform complex manipulation tasks with sensitivity and accuracy, akin to human touch (Shirafuji & Hosoda, 2011).

ACTUATORS IN SOCIAL ROBOTIC SYSTEMS

Actuators are typically devices that transform different forms of energy (e.g., electrical, hydraulic, pneumatic) into physical movement. Actuators can produce various types of motion, including linear motion, rotational motion, oscillatory motion, and combinations thereof, depending on their design and application. After energy is input, the conversion is facilitated by the components of the actuator such as motors, pistons, gears, or solenoid electromagnets. Actuators are typically classified into different categories based on their operating principles (Ogenyi, 2019). The choice of the actuation technology to use depends on the design requirements and intended functionalities of the robot; each of these actuators comes with some strengths and weaknesses.

Pneumatic Actuators

Pneumatic actuators are those that utilize compressed air to generate motion. Specifically, they rely on a piston and cylinder air-compressing system which generate pressure by way of air displacement (Robinson, 2014). This pressure propels the piston, inducing mechanical motion as an output. This kind of actuator is desirable for a process that require quick and precise movements because by regulating air pressure, the extent and timing of motion can be finely tuned, enabling accurate positioning and operation, and for a process in a safety-conscious robotic design, especially, those intended to interact with human and delicate contact surfaces as they do not use flammable fluids and are less prone to leakage (Ogenyi, 2019). This kind of actuator has gained applications in various fields including in creating artificial muscle for robot joints, which allow for generation of mechanical motion (Tang & Gong, 2023).

Electromechanical Actuators

Electromechanical actuators are those that use electric current or magnetic field to bring about a physical motion. The common energy conversion facilitators of this set of actuators are the electric motors that convert electrical energy into rotational or linear motion. The motors commonly used in robotics come in various types, each offering unique characteristics suited for different applications and requirements. The servo motors come with built-in feedback mechanisms such as encoders (i.e., a sensing device that provides feedback on motion, position, or direction) that helps for precise control of rotary or linear actuation suitable for robot joint motion (Hirose & Ogawa, 2007). Similarly, the stepper motor operates by dividing the full rotary motion into a series of steps which is achieved by energizing one or more of the stator phases (i.e., number of independent coils) arranged around a rotor. By energizing these coils in a specific sequence, the motor can rotate step by step (Gautam et al., 2017). Stepper motors are implemented when precise positioning or precise speed control or both are required in automation systems. Another electromechanical actuator presently used in social robots is the linear actuator motor which convert rotary motion into linear motion thereby providing high-speed and accurate positioning of robot joints (Maziz et al., 2015).

Hydraulic Actuators

Hydraulic actuators are those that rely on compressed fluid for motion generation. They generate motion through the coordination of a cylinder containing hydraulic fluid and a piston. The process involves pressurizing the hydraulic fluid through a pump, which causes the piston to move linearly within the cylinder. This linear motion is subsequently transformed into the desired movement, such as the articulation of limbs or joints in social robots. The hydraulic actuator is good for carrying heavy loads, because it operates at high pressures, which allows it to exert significant force on the load, therefore, enabling them to move heavy objects by overcoming resistance. It is generally known to have fewer problems when exposed to heat compared to pneumatic actuators due to the incompressibility of hydraulic fluids, which helps maintain consistent performance even in high-temperature environments. However, the hydraulic actuators could be susceptible to the temperature of the liquid used. This makes the actuators difficult to miniaturize because the viscosity of hydraulic fluids can change with temperature variations, affecting the performance of hydraulic actuators making them difficult to be used in smaller robots (Ogenyi, 2019).

SENSOR-ACTUATOR COORDINATION IN EVERYDAY CONTEXTS

Given the role of sensor-actuator synergies in the basic functioning of social robots, their importance to how robots will function in everyday contexts cannot be understated. In particular, these components will be integral to robot functioning in smart cities, which are urban areas that use data-driven solutions to enhance the quality of life for its inhabitants, improve sustainability, and optimize resource efficiency. Robots are expected to play a key role in smart cities as they move their machinic bodies through traditionally human spaces like streets, offices, commercial locales, and even personal spaces (Macrorie et al., 2021). For instance, robots equipped with sensors like cameras can patrol and monitor a smart city to detect unusual activities or security threats (Mohamed et al., 2020). Through the Internet of Things (IoT, or networks of sensing objects), this information can be shared among other robotic systems within the city for collaborative support. In the event of security breaches, these robots can use their equipped actuators to deploy barriers and alarms for swift response.

Moreover, the synergy between sensors and actuators enables robots to monitor the environmental health status of smart cities (Studley & Little, 2021). Sensors can detect and record crucial environmental parameters such as air quality, temperature, and pollution levels (Rolddan-Gomez et al., 2022). Actuators play a vital role in this scenario by enabling robots to take immediate action. They can deploy remediation measures such as spraying water to reduce air pollution, releasing chemicals to neutralize contaminants, or adjusting valves to control water flow in drainage systems (see Yu, 2014). This integrated approach enhances the effectiveness of environmental monitoring and management in smart cities (Yu, 2014).

Although the sensor-actuator synergy in robotics offers numerous benefits for smart cities, it also raises ethical concerns regarding data privacy and security (Ahmad et al., 2022). For instance, there's a risk that sensitive information captured by robotic systems, such as personal identity data, could fall into the wrong hands and be misused. Additionally, in regions with lax data protection regulations, individuals may have limited control over how their data is stored and used by authorities or third parties (Yaacoub et al., 2022). For example, consider a scenario where surveillance robots equipped with facial recognition technology patrol public areas to enhance security. While this can help identify potential threats, there's a risk that the collected facial data could be misused for unauthorized surveillance or profiling purposes. Moreover, if this data is not adequately protected, it could be vulnerable to hacking or data breaches, leading to privacy violations and identity theft.

Therefore, it's essential for policymakers, technology developers, and society as a whole to address these ethical concerns and implement robust measures to safeguard data privacy and security in the context of sensor-actuator-enabled robotic systems deployed in smart cities. This may include enacting stringent data protection laws, implementing encryption and authentication mechanisms, and promoting transparency and accountability in data handling practices.

SENSATION BEYOND THE HUMAN

In considering how social robots may function in human spaces, we must also consider the way that robots' machine capabilities go *beyond* human norms for perception. Beyond the visual, audio, proximity, and tactile sensors mentioned above, robots can have other types of sensors that carry implications for social environments.

Biometric sensors are devices that catch and analyze biological attributes or behavioral characteristics of individual, for example fingerprints, facial elements, iris patterns, voiceprints, or stride, with the end goal of validation, distinguishing proof, or observing. The term "validation" refers to confirming the identity of an individual, "proof" pertains to verifying their identity or characteristics, and "observing" refers to monitoring their behavior or activities for various purposes, such as security or personalized services. This sensor works by capturing biometric traits of individuals and converting them into digital format that can be compared against stored or reference data to determine if there is a match. This could find applications in personalized, real-time human-robot interaction (Foggia et al., 2023). For example, the biometric sensor could be used in retail or hospitality settings where a robot customer service agent can use facial recognition to greet customers by name, remember past interactions, and provide tailored assistance or recommendations.

Gyroscopes are sensors used to measure orientation, such as tilting or rotation, and are crucial for stabilizing and controlling the orientation of devices like drones, aircraft, and robots (Li et al., 2016). Gyroscopes utilize spinning masses to resist changes in orientation and exhibit gyroscopic precession when subjected to external forces. They are essential components of humanoid robots as they help maintain balance on the floor. Accelerometers are sensors used to determine linear acceleration of a target object along the XYZ coordinates. They can detect sudden changes in acceleration, enabling robots to identify potential falls and take preventive actions.

Now, these sensing and movement abilities are in *some ways* like those of humans. We can certainly recognize faces and voices, though we cannot do it at a grand scale, do it perfectly, or even necessarily do it reliably. We can balance and rotate on our feet atop shifting ground, but we are clumsy and fallible in that effort. Others among these are likely beyond our capabilities, as with the capture, remembering, and retrieval of fingerprints or iris patterns. But perhaps the most interesting way that robot sensing and motion go beyond our own is in their integration. In most cases, the information from a single sensor would not be enough to provide a robot with the needed information to effectively interact with a human collaborator. This is similar to our human need to see and hear and feel and smell in order to have a rich understanding of our environments and our interactions therein. For robots, this requirement led to the birth of *sensor fusion*. Sensor fusion is the algorithmic integration of multiple sensory inputs, to improve the accuracy, reliability, precision, robustness, adaptability, and enhance the overall performance of the robot (Shen et al., 2021). Notably, humans also integrate sights, sounds, and smells, so an open question remains—what does the integration of those beyond-human sensations mean for human-robot interactions and especially the social power that comes with insight about others? In other words, when robots can reliability and perhaps perfectly remember our identities and tendencies and also outmaneuver us in physical activities, what could this mean for how robots could hold social or physical power as we interact in everyday life?

REFERENCES

Ahmad, K., Maabreh, M., Ghaly, M., Khan, K., Qadir, J., & Al-Fuqaha, A. (2022). Developing future human-centered smart cities: Critical analysis of smart city security, data management, and ethical challenges. *Computer Science Review, 43*, 100452.

ClearPath Robotics. (n.d.). Turtlebot 4. <https://clearpathrobotics.com/turtlebot-4/>

Fiore, S. M. (2013). Toward understanding social cues and signals in human-robot interaction: Effects of robot gaze and proxemic behavior. *Frontiers in Psychology, 4*, 859.

Foggia, P., Greco, A., Roberto, A., Saggese, A., & Vento, M. (2023). A social robot architecture for personalized real-time human-robot interaction. *IEEE Internet of Things Journal, 10*(24), 22427–22439.

Gautam, R., Gedam, A., Zade, A., & Mahawadiwar, A. (2017). Review on development of industrial robotic arm. *International Research Journal of Engineering and Technology, 4*(3), 1752–1755.

Hirose, M., & Ogawa, K. (2007). Honda humanoid robots development. *Philosophical Transactions of the Royal Society A: Mathematical, Physical and Engineering Sciences, 365*, 11–19.

Kappassov, Z., Corrales, J., & Perdereau, V. (2015). Tactile sensing in dexterous robot hands. *Robotics and Autonomous Systems, 74*, 195–220.

Lee, J., & Ahn, B. (2020). Real-time human action recognition with a low-cost RGB camera and mobile robot platform. *Sensors, 20*, 2886.

Li, B. Boccanfuse, L., Wang, Q., Barney, E., Ahn, Y. A., Foster, C., Chawarska, K., Scassellati, B., & Shic, F. (2016). Human robot activity classification based on accelerometer and gyroscope. In *Proceedings of the International symposium on Robot and Human Interactive Communication* (pp. 423–424). IEEE.

Li, J., Mi, Y., & Li, G. (2019). CNN-based facial expression recognition from annotated RGB-D images for human-robot interaction. *International Journal of Humanoid Robotics, 16*(4), 1941002.

Macrorie, R. Marvin, S., & While, A. (2021). Robotics and automation in the city: A research agenda. *Urban Geography, 42*, 197–217.

Marge, M., Espy-Wilson, C., Ward, N., Alwan, A., Artizi, Y., Bansal, M., … Yu, Z. (2022). Spoken language interaction with robots: Recommendations for future research. *Computer Speech & Language, 71*, 101255.

Maziz, A., Alexandre, K., Nils-Krister, P., & Edwin, J. (2015). Soft linear electroactive polymer actuators based on polypyrrole. In *Proceedings of Electroactive Polymer Actuators and Devices* (pp. 289–294). SPIE.

Mohamed, N., Al-Jaroodi, J., Jawhar, I., Idries, A., & Mohammed, F. (2020). Unmanned aerial vehicles applications in future smart cities. *Technological Forecasting and Social Change, 153*, 119293.

Moller, R., Furnari, A., Battiato, S., Härmä, A., & Farinella, G. M. (2021). A survey on human-aware robot navigation. *Robotics and Autonomous Systems, 145*, 103837.

Nguyen, L. A., Thang, D. M., Dung, P. T., Khoa, T. D., Son, N. H., Hiep, N. T., Nguyen, P. V., Truong, V. D., Toan, D. H., Hung, N. M., Ngo, T., & Truong, X. (2018). Deep learning-based multiple objects detection and tracking system for socially aware mobile robot navigation framework. *Proceedings of the NAFOSTED Conference on Information and Computer Science* (pp. 436–441). IEEE.

Ogenyi, U. E. (2019). Physical human-robot collaboration: Robotic systems, learning methods, collaborative strategies, sensors, and actuators. *IEEE Transactions on Cybernetics, 51*, 1888–1901.

Rascon, C. & Meza I. (2017). Localization of sound sources in robotics: A review. *Robotics and Autonomous Systems, 96*, 184–210.

Robinson, R. M. (2014). *Pneumatic artificial muscle actuators for compliant robotic manipulators* (Publication No. AAT 3644408) [Doctoral dissertation, University of Maryland, College Park]. ProQuest Dissertations & Theses Global.

Rolddan-Gomez, J. J., Garcia-Aunon, P., Mazariegos, P., & Barrientos, A. (2022). SwarmCity project: Monitoring traffic, pedestrians, climate, and pollution with an aerial robotic swarm: Data collection and fusion in a smart city, and its representation using virtual reality. *Personal and Ubiquitous Computing, 26*, 1151–1167.

Shen, Z., Elibol, A., & Chong, N. Y. (2021). Multi-modal feature fusion for better understanding of human personality traits in social human-robot interaction. *Robotics and Autonomous Systems, 146*, 103874.

Shirafuji, S., & Hosoda, K. (2011). Detection and prevention of slip using sensors with different properties embedded in elastic artificial skin on the basis of previous experience. In *Proceedings of the International Conference on Advanced Robotics* (pp. 459–464). IEEE.

Studley, M. E., & Little, H. (2021). Robots in smart cities. In M. I. A. Ferreira (Ed.), *How smart is your city? Technological innovation, ethics and inclusiveness* (pp. 75–88). Springer.

Tang, Y., & Gong, D. (2023). Bio-inspired cooperative control for robotic joint actuated via sequential recruitment of multiple pneumatic artificial muscles. In *Proceedings of the Chinese Control and Decision Conference* (pp. 3286–3291). IEEE.

Tashtoush, T., Garcia, L., Landa, G., Amor, F., Laborde, A. N., Oliva, D., & Safar, F. (2021). Human-robot interaction and collaboration (HRI-c) utilizing top-view RGB-D camera system. *International Journal of Advanced Computer Science and Applications, 12*(1), 10–17.

Yaacoub, J.-P. A., Noura, H. N., Salman, O., & Chehab, A. (2022). Robotics cyber security: Vulnerabilities, attacks, countermeasures, and recommendations. *International Journal of Information Security, 21*, 115–158.

Yu, S. (2014). Water spray geoengineering to clean air pollution for mitigating haze in China's cities. *Environmental Chemistry Letters, 12*(1), 109–116.

CHAPTER THIRTEEN

Implants & Injections: Long-Lived Integrations with the Organic

DAYEOUN JANG & STEPHANIE JORDAN

Social robots redefine humanity. Indeed, the category of *human beings* who can enjoy *human rights* has been defined differently by times, cultures, and hegemonies (Harvey, 1994). Now, social robots bring the next revision. The development of social robots forces us to revisit the essential meaning of being human, including criteria for that category and what an acceptable normal body is. This chapter devotes discussion to understanding the future that social robots bring as they help to inform our considerations of cyborgs. The ontological categories of social robots and cyborgs are not equitable and rather overlapping. Even if social robots are not allowed to have warmth, flesh, and pulsating organic bodies like cyborgs, cyborgs and social robots are partially or wholly borne from non-organic mechanical parts. We argue that we can understand cyborgs through understanding social robots and vice versa.

The word "cyborg" is a combination of two words: Cybernetics and organism (Clynes & Kline, 1960). Between the two words, the word *organism* is quite clear—living things—but cybernetics is a little bit tricky. Cybernetics refers to all entities, including animals and machines, that exist inside of control and communication systems (Wiener, 1961/2019). Putting the two together, cyborgs are a type of "cybernetically extended organism" (Kline, 2009, p. 332). By artificial connection with cybernetics, an organism can extend, improve, adjust, or normalize its self-regulatory system (Clynes & Kline, 1960), and this extension process always requires continuous negotiation between the organismic and cybernetic components. In this regard, cyborgs always exist in unstable, fluid, and forever-changing dynamics.

In this chapter, cybernetic technology (CT) refers to a communicative technology or non-organic object that controls humans or is controlled by humans. Based on current technological development trends, CTs can include implants and injections. The former are technological devices implanted in a body, and the latter are technological devices injected through needles. There are also ingestible technologies like supplements and medicines that can be considered CTs (Barfield & Williams, 2017). Even if CTs each have different purposes and appearance, most CTs share one characteristic: Irreversibility. Once a CT is implanted or injected, detaching it from a body is often impossible or requires extreme effort and costs. Some of the most common CTs are for medical or aesthetic purposes, such as pacemakers, limb prostheses, cochlear implants, hormones, insulin, facial fillers, or Botox. These examples typically serve the function of bringing a person a normal or better life.

There is another application of CTs, characterized in the negative by Jillian Weise (2016) as the *tryborg*, which are non-essential enhancements and extensions. A common example of the tryborg includes body hackers who implant RFIDs or microchips under their skin for convenience purposes and to replace IDs and door keys. Whether the intervention of CT is essential or not, the existence of cyborgs and tryborgs underlie key issues in science and technology studies (STS) of robots that ask how individuals accrue authenticity, authority, and autonomy under the growing entanglement of emerging technologies (Hancock, 2020). Artist and technologist Neil Harbisson, for instance, is a visual artist who is colorblind; Harbisson implanted an antenna in his skull that can turn light frequencies into sounds to feel colors (Barfield & Williams, 2017). In concert with Harbisson, avant-garde artist and dancer Moon Ribas implanted a vibrating device in her feet and torso to internalize the sensation of earthquakes around the world and attempt to embody climate change as earthquakes increase over time (Guler et al., 2016).

Their collaboration is not solely artistic but also political. A body is a place where political debates are continuously progressing. A human body cannot exist or be constructed outside of society; society regulates a body and, at the same time, an individual resists the power of society to express a unique identity and have agency and authority (Bordo, 1994). As society is not fixed (Elder, 1994), the politics of the body are always changing in a way of refusing, adapting, negotiating, or conforming with society. Given that, Harbisson and Ribas bring an additional agent—technology—to the politics of the body. Until the CTs are removed from the body, they will be another participant in the evolving debates on the politics of the body.

Now, more cyborgian beings are emerging and becoming visible, although it is largely limited to CTs for medical purposes. For example, there are an increasing number of TikTok or YouTube creators with limb differences or cochlear implants, and they are socially accepted and normalized from fashion shows' runway to daily life. This highlights a modern interpretation of "the cyborg's dilemma" (Biocca, 1997, p. 1), which refers to a paradox that as technology increasingly mimics natural human interactions, it simultaneously alters these very interactions. CTs have penetrated daily life and changed the way humans perceive and interact with machines that imitate humans or a part of human bodies. We note that this intervention does not happen immediately and temporarily; it does slowly and gradually by building a long-term relationship with humans, just as social robots will do.

We frame cyborgs in terms of ongoing relationship-building with non-organic objects and their maintenance through long-term shifts in societal and professional norms and cultural change, where dominant political and ideological trends dictate what technology means to humanity's ability to sustain and survive.

WE ARE ALL CYBORGS

Some scholars suggest humans have always been cyborgs (e.g., Sorgner, 2021). Understanding humans as cyborgs means that natural bodies are always in juxtaposition with the technologies around them through intervention, extension, and enhancement outside of any innate organic matter. Technologies are built within ideological frames and often reify the values of their designers (Miller, 2021), which blends with the biological and social function of the human being that carries it.

The ideologies that define cyborgian beings have changed. The original definition of cyborg refers to a system that deliberately incorporates exogenous components extending the function of human or other living organisms (such as a mouse with an external osmotic pump from Clynes and Kline's original paper; 1960) in order to adapt to new space colonies. Cyborgs, then, are new possibilities and the promise of an enhanced future. Interestingly, even before the term cyborg emerged, the imagination and possibility of human entanglement with technology existed. However, most depictions of cyborgs were related to the monstrous, abnormal, and terrifying characters appearing in horror and mystery genres, such as novels like *Frankenstein* (Shelley, 1818/2012) and films like *Doctor X* (Curtiz, 1932).

Early and recent media portrayals depict cyborgs aligned with negative imaginations of technology (e.g., losing one's essence or autonomy or being absorbed into a collective mass). Of course, there could be multiple reasons for the biased representation, but one interesting and valid hypothesis is cyborgs are a metaphor for technophobia (Jones, 2021). Horror movies reflect society's fear of or desire for a certain subject (Hendershot, 2001), and cyborgs are a good strategy to deepen the uncanny by "reminding [readers] of the inextricability of the machine and the human" (Punter, 2012, p. 263). Indeed, cyborgs symbolize "contemporary versions of monsters" (Rayner, 1994, p. 126) as pagans did in medieval Europe, and cyborgs represent fears and complex relationships with technology.

Such monsters have served important social and psychological functions, especially in building individual identities and predominant ideologies. People and society define themselves by defining strangers as worshipped gods or excluded monsters and then by comparing themselves with those two (Kearney, 2003). Strangers are unnatural beings that reside outside of our epistemology (Rayner, 1994), so they always become Others. Cyborgs are also strangers in society; they exist outside of the natural epistemology and cannot be categorized into either humans or technologies (Haddow et al., 2015). Therefore, society treats cyborgs as monsters or gods; they can never be accepted in society, unlike acceptable "normal" people and technology.

Donna Haraway (1987), however, has grappled with the monster-versus-god narrativization. She criticizes goddess feminism, which refuses masculine technology and advocates for a naturalistic feminist divinity. She rejects the notion that "natural" means "normal" by canonically asserting, "I'd rather be a cyborg than a goddess" (p. 37). Instead, she asserts that we must embrace strangers, and she dreams of a utopia free from all different forms of Othering and categorizing. To her, cyborgs are symbols of revolution by blurring existing binaries, rejecting the notion of organismal wholeness, and favoring a complexified identity through union with non-humans.

LONG-LIVED IDEOLOGIES

How one divides acceptable normal entities from strangers varies depending on one's standpoint (Harding, 2004), and is usually decided by ideological frames (e.g., religion, political affiliation) and social systems (e.g., cultural traditions, group norms). Among those, religious doctrines have long had the power to separate normal familiars and abnormal strangers; they set up criteria that should be satisfied to be accepted (Holman, 2012; Rayner, 1994), and the body is important. Based on their concept of the body bestowed by a god, they enact rules about the body (Gunderman & Jackson, 2013). For example, Abrahamic religions (e.g., Judaism, Christianity, Islam) and Confucianism have similarly conservative stances on body modification and even on related behaviors such as cutting hair, getting tattoos, or eating constituting foods (e.g., Holman, 2012).

These ideologies also have a strong influence on how society and individuals perceive the technology surrounding bodies. For instance, since the late 2000s, when biologically implantable microchips became visible to the public, some

Christian-based organizations have claimed that these implanted chips represent the mark of the beast prophesied in the Book of Revelation (Foster & Jaeger, 2008). The negative perception of these implants, combined with the recent fake news that the coronavirus vaccine contained a microchip, has led some people to reject the vaccine (Dwoskin, 2021).

In turn, Buddhism has no fixed concept of a body and often endorses body modification as a means of eliminating suffering and freeing from the obsession with physical and material forms (Hughes, 2019). The middle way, one of the main concepts of Buddhism, emphasizes not being constrained by the extremes; there is no difference between extremes because every element in the world is connected (Hall, 2006), and the two extremes of machines and humans are the same. It seems natural, then, that these core Buddhist beliefs are positively influencing how people view cyborgs in modern times. A recent study has revealed that Asian nations with deep-rooted Buddhist heritage (e.g., Thailand, Cambodia) exhibit more favorable attitudes toward body modifications, especially gender-affirming care (Macer, 2012).

The positive portrayal of cyborgs in media is also sometimes related to Buddhism. In the game *Overwatch*, for example, the character Genji replaced most of his body parts with machines after being defeated and severely injured. He was in agony from shame of his mechanical body but accepted and became proud of his cyborgian identity after guidance from Zenyatta, who is a Buddhist monk robot. Although we cannot know how Zenyatta helped Genji find inner peace, Zenyatta's philosophy represents the middle way; there is no difference between humans and robots. Thus, the ideological standing point is essential and crucial in evaluating whether the meaning and relationship between technology is based on robust separation or harmonious enlightenment.

THE ACCEPTABILITY POLITICS OF THE CYBORG

Ideologies and social systems not only limit the acceptable, normal body; they also shape notions of authenticity and integrity—and the consequences thereof. Most CTs, mostly implants and injections, are invisible and hard to detect by the bare eye: Doping and steroids, hormones, Tommy John surgery, and implantable cardioverter defibrillators. Instead, it is only visible in testing or in one's ability and capacity to perform different functions without harm. Because we cannot easily recognize those invisible cyborgs, we cannot clearly know one's competitive advantages originate from their body or CTs. This brings another fear and anxiety in society: Unfairness.

Especially in sports, where the human body becomes perfectible and inspectable, CTs have been not only accepted but embraced (Fouché, 2012) because they are testable, quantifiable, and useful in enhancing performance. However, not all cyborgs are accepted and embraced, especially when a CT is too successful and visible. Consider Oscar Pistorius. His J-shaped carbon prosthetic leg received significant international attention because it made him break the limit of a disabled body. However, his prosthetic leg was banned under the new rule, which prohibits the use of mechanical devices, giving them competitive advantages. Fouché (2017) argues that the rule is based on the concerns and worries that athletes with a disability may harm the Olympics' ideal, which pursues "able-bodied physical perfection and performance" (p. 112).

Pistorius' cases exemplify the considerable effects of ideologies and systems in judgements of cyborg acceptability politics. Rules, norms, and thresholds dictate what is fair and unfair and what is appropriate and inappropriate for an organic or augmented body participating in competitions. Against the dictatorship, cyborgs demonstrate that there is no actual and solid boundary between acceptable normalcy and unacceptable abnormality, by bringing forward uncomfortable dialogs about performance, ability, integrity, authenticity, and the tenuous line between ability and disability.

BROKEN-WORLD THINKING: LONG-LIVED CYBORGS

Cyborgs' lifetime relationships with entangled technologies are fluid rather than fixed. Cyborg bodies are subject to reconfiguration to fit within the ideological purview of the social world in which they live, animating ongoing material and relational maintenance. For example, Cochlear implants are often considered a miracle, but not many people know that a few years of training processes are required to adapt to them (Mauldin, 2019). Throughout the process, the expectation of the miracle will disappear, but they set a new expectation based on constant negotiation with the treatment. Besides, Schuster (2021) shows the lifetime of medical treatment that defines gender-affirming care through implants and injections. Over time, every aspect of that care is shifting, including norms, dispositions of individual medical personnel, insurance policies, identities, and even technological capabilities.

Likewise, the meaning of cyborgs and social robots is always changing based on surrounding contexts. In imagining the uncertain relationships that will be uncovered as humans and nonhumans enmesh in new ways over time, we argue the importance of "broken-world thinking" (Jackson, 2014, p. 221), highlighting the complementary relationship between humans and technology in everyday life. Jackson states that "the world is always breaking; it's in its nature to break" (p. 223). However, we are rarely aware that the world is breaking because infrastructure is only visible upon breakdown (Star, 1999). By then, we can realize the fragile and messy nature of technology covered by well-organized systems. For example, we usually do not think about how our smartphone is connected to the Internet because they work without problems. The existence of the system will be faded once we are indifferent to it. However, when the Internet connection is lost, we feel the system and see the people who maintain it. Given that cyborgs and social robots tear apart the definition of humans and reveal the vulnerability and imperfection of bodies, cyborgs and social robots are essentially based on broken-world thinking—they expose the hidden dynamics of humans and technology based on constant material and relational maintenance.

Cyborgs' broken-world thinking can have a greater impact in terms of complex entanglements between bodies and CTs. Once CTs are implemented into bodies, they will not automatically integrate into bodies. Instead, they will underscore the imperfection and unstable connection with them. The old way of maintaining bodies is no longer valid, and they must learn how to fix technical errors and define themselves with CTs. We call this *cyborgian thinking*, which is based on broken-world thinking but emphasizes the personal and intimate relationships between humans and CTs. Cyborgian thinking includes realizing the susceptibility of the relationships between a human body and CTs and, most importantly, symbiosis with CTs based on the endless engagement in maintenance.

Jackson (2014) asks us how we might build new and different forms of solidarity with non-human objects in a constantly changing, broken world. In the changing dynamics, the emphasis on symbiotic relationships between humans and CTs allows humans to find a new way to define their identities with CTs. For example, people may decorate their prosthetic legs or arms under the name of fashion. You can easily find Pinterest posts that collect brilliant ideas to decorate a prosthetic limb, and the market for prosthetic limb covers is increasing rapidly, and there is a jewelry brand for prosthetic limbs. We can find a similar example from recent history: Eyeglasses. The meaning and the relationship with eyeglasses have expanded from corrective purpose to fashion for centuries (Pullin, 2011), and prosthetic limbs may have their own value in expressing unique identities.

Long-lived relations between technology and humanity are understudied and undertheorized. Still, attention to maintenance can help us better understand how technology and humans have influenced each other in a symbiotic relationship. In a similar shift, we can understand the ongoing underlying dynamics that define our relationship with social robots and CTs as similarly fragile and evolving.

HUMAN, CYBORG, AND BEYOND: IMPLICATIONS FOR ROBOTS

If social robots are to co-construct life with humans, then any assertion of monstrousness or divinity previously assigned to cyborgs becomes ill-fit. The current moment is instead defined by humans' deep need and even affection for technologies that aid in their everyday behaviors, from survival and interactions to work and even mingling. Therefore, much as we have demonstrated that ideologies change over time, we emphasize how the definition and relationship between humans and robots have already changed and require integration into how we imagine our futures. Cyborgian thinking—that is, ideation that engages the broken-world premise based on our body—deconstructs the dichotomy between human and robot or divine and monster and, instead, forces people to redefine humanity in the age of social robots.

We argue in this chapter that the notion of the normal and acceptable body that society suggests and many of us take for granted is nothing more than a product of ideology, time, and circumstance. The concept of the cyborg makes us consider the vulnerability and maturation of the human body, the need for repair of technologies over time, and the requirement and criticality for understanding the ongoing symbiotic relationships between humans and nonhumans.

This ongoing, long-lived relationship extends the productivist drive that dominates technological discourse and the innovation speak of design worlds. Instead, we must redefine our commitments: Ongoing adaptation and revision, knowledge about how the body ages and incorporates new material and social relationships, and emphasis on the care and maintenance of relationships and materials. By depicting the nuanced complexities of the human-machine relationship as it shifts temporally, ideologically, and materially, what is most intimate to us in an age of social robots provokes a new definition of humanity and survival. By understanding the cyborgs that already exist in our societies, we are rewriting the definition of human and robot as one.

REFERENCES

Barfield, W., & Williams, A. (2017). Cyborgs and enhancement technology. *Philosophies, 2*(1), 4.
Biocca, F. (1997). The cyborg's dilemma: Progressive embodiment in virtual environments. *Journal of Computer-Mediated Communication, 3*(2), JCMC324.
Bordo, S. (1994). Feminism, Foucault and the politics of the body. In R. Miguel-Alfonso & S. Caporale-Bizzini (Eds.), *Reconstructing Foucault* (pp. 219–243). Brill.
Clynes, M. E., & Kline, N. S. (1960). Cyborgs and space. *Astronautics, 14*(9), 26–27.
Curtiz, M. (Director). (1932). *Doctor X* [Film]. First National Pictures.

Dwoskin, E. (2021). On social media, vaccine misinformation mixes with extreme faith. *The Washington Post*. Retrieved from <https://www.washingtonpost.com/technology/2021/02/16/covid-vaccine-misinformation-evangelical-mark-beast/>

Elder, G. H. (1994). Time, human agency, and social change: Perspectives on the life course. *Social Psychology Quarterly, 57*(1), 4–15.

Foster, K. R., & Jaeger, J. (2008). Ethical implications of implantable radiofrequency identification (RFID) tags in humans. *The American Journal of Bioethics, 8*(8), 44–48.

Fouché, R. (2012). Aren't athletes cyborgs?: Technology, bodies, and sporting competitions. *Women's Studies Quarterly, 40*(1/2), 281–293.

Fouché, R. (2017). *Game changer: The technoscientific revolution in sports*. JHU Press.

Guler, S. D., Gannon, M., & Sicchio, K. (2016). Superhumans and cyborgs. In S. D. Guler, M. Gannon, & K. Sicchio (Eds.), *Crafting wearables: Blending technology with fashion* (pp. 145–159). Apress.

Gunderman, R. B., & Jackson, C. R. (2013). Viewing the body in the Abrahamic faith traditions: What radiologists need to know. *Academic Radiology, 20*(4), 506–508.

Haddow, G., King, E., Kunkler, I., & McLaren, D. (2015). Cyborgs in the everyday: Masculinity and biosensing prostate cancer. *Science as Culture, 24*(4), 484–506.

Hall, C. M. (2006). Buddhism, tourism, and the middle way. In D. Timothy, & D. Olsen (Eds.), *Tourism, religion and spiritual journeys* (pp. 172–185). Routledge.

Hancock, E. (2020). Should society accept sex robots? Changing my perspective on sex robots through researching the future of intimacy. *Paladyn, Journal of Behavioral Robotics, 11*(1), 428–442.

Haraway, D. (1987). A manifesto for cyborgs: Science, technology, and socialist feminism in the 1980s. *Australian Feminist Studies, 2*(4), 1–42.

Harding, S. G. (2004). *The Feminist Standpoint Theory reader: Intellectual and political controversies*. Psychology Press.

Harvey, L. S. C. (1994). Mr. Jefferson's wolf: Slavery and the suburban robot. *Journal of American Culture, 17*(4), 79–79.

Hendershot, C. (2001). *I was a cold war monster: Horror films, eroticism, and the cold war imagination*. Popular Press.

Holman, A. (2012). Religion and the body an overview of the insertions of religion in the empirical psycho-social research lines on the body. *European Journal of Science and Theology, 8*(3), 127–134.

Hughes, J. J. (2019). Buddhism and our posthuman future. *Sophia, 58*(4), 653–662.

Jackson, S. J. (2014). Rethinking repair. In T. Gillespie, P. J. Boczkowski, & K. A. Foot (Eds.), *Media technologies: Essays on communication, materiality, and society* (pp. 221–239). MIT Press.

Jones, D. (2021). *Horror: A very short introduction*. Oxford University Press.

Kearney, R. (2003). *Strangers, gods, and monsters: Interpreting otherness*. Psychology Press.

Kline, R. (2009). Where are the cyborgs in cybernetics? *Social Studies of Science, 39*(3), 331–362.

Macer, D. (2012). Ethical consequences of the positive views of enhancement in Asia. *Health Care Analysis, 20*(4), 385–397.

Mauldin, L. (2019). Don't look at it as a miracle cure: Contested notions of success and failure in family narratives of pediatric cochlear implantation. *Social Science & Medicine, 228*, 117–125.

Miller, B. (2021). Is technology value-neutral? *Science, Technology & Human Values, 46*(1), 53–80.

Pullin, G. (2011). *Design meets disability*. MIT Press.

Punter, D. (2012). Shape and shadow: On poetry and the uncanny. In D. Punter (Ed.), *A new companion to the Gothic* (pp. 252–264). Wiley-Blackwell.

Rayner, A. (1994). Cyborgs and replicants: On the boundaries. *Discourse, 16*(3), 124–143.

Schuster, S. (2021). *Trans medicine: The emergence and practice of treating gender*. New York University Press.

Shelley, M. W. (2012). *Frankenstein; or, The modern Prometheus*. Project Gutenberg. Retrieved from <https://www.gutenberg.org/ebooks/41445> (Original work published 1818)

Sorgner, S. L. (2021). *We have always been cyborgs: Digital data, gene technologies, and an ethics of transhumanism*. Policy Press.

Star, S. L. (1999). The ethnography of infrastructure. *The American Behavioral Scientist, 43*(3), 377–391.

Weise, J. (2016). The dawn of the Tryborg. *The New York Times*. Retrieved from <https://www.nytimes.com/2016/11/30/opinion/the-dawn-of-the-tryborg.html>

Wiener, N. (with Hill, D., & Mitter, S.). (2019). *Cybernetics or control and communication in the animal and the machine, Reissue of the 1961 Second Edition*. MIT Press. (Original work published 1961)

CHAPTER FOURTEEN

Aggregation & Distribution: Beyond the Singular Form

SARAH DIEFENBACH, DANIEL ULLRICH, & ANDREAS BUTZ[1]

Though the definitions vary in detail, most agree that a robot is a machine capable of carrying out physical tasks automatically, whereby social robots are robots specifically designed for interaction with humans. Social robots, thus, can be seen as a subcategory of assistive robots and non-industrial robots, with the purpose of physical or social assistance or companionship (Heerink et al., 2010). In many cases, social robots are designed with anthropomorphic shapes (see Chapter One) and the interaction builds on patterns from interaction between humans (e.g., Bartneck & Forlizzi, 2004), such as rebuilding human social norms and communication via speech. A typical application in everyday settings would be a shop assistant robot. For example, KeJia is a 165cm tall service robot applied in a large shopping mall in Hefei city, China (Chen et al., 2017). KeJia has an anthropomorphic appearance and is dressed in a traditional Chinese suit, which shall encourage people to interact with it in a natural way (Chen et al., 2017, p. 4). KeJia is equipped with camera, microphone and speech synthesis module, and the whole robot is motorized by two differential wheels. KeJia's arms are flexible and can make some gestures, like greeting people or pointing directions. If a customer needs help, the customer can speak to the robot—and if the robot

---

1 Part of this research was funded by the DFG-project PerforM—Personalities for Machinery in Personal Pervasive Smart Spaces (DI 2436/2-1) in context of the Priority Program "Scalable Interaction Paradigms for Pervasive Computing Environments" (SPP 2199).

understands what the customer is looking for, the robot directs the customer to the right location.

In many respects, this type of interaction resembles what humans are used to when interacting with other humans. Usually, all the entities that we interact with have a body. But this stereotypical thinking about robots with singular bodies limits our imagination of what a social robot could be. What if, besides a singular body, there is a degree of distribution that robots could have?

Distribution means that single parts of an entity are spread out over an area or throughout a space or unit of time. Though in daily life, we are not used to social interaction with distributed entities, there are already examples in movies, like Keanu Reeves playing Neo and meeting Deus ex machina in the film *The Matrix Revolutions* (Wachowski & Wachowski, 2003), or the family movie *Big Bird in China* (Stone, 1983). In the latter movie, Big Bird and his dog Barkley from Sesame Street travel to China to search for the Phoenix Bird, which they saw pictured on a treasure map they found in a Chinese store. During their travel through China, they solve many puzzles and mysteries and learn a lot about the country, its people and culture. However, though they found all the spots pictured on the treasure map, they still wait for the Phoenix Bird and its message, curious to hear what a Chinese bird will have to tell an American bird. They are almost about to give up, when suddenly the Phoenix Bird shows up in a tree. However, it is not a bird with a body like Big Bird but appears as a sum of light dots. In the beginning, the light dots look more like a shooting star, then form the shape of a bird, but also keep moving which looks like the Phoenix is spreading its wings. The Phoenix Bird's appearance is distributed but still clearly recognizable as forming one entity and becoming recognizable as interaction partner to Big Bird. Then Big Bird speaks to the Phoenix and asks for its message. The Phoenix, however, answers that they already got the message, since what Big Bird and Barkley have experienced during their travel is more wisdom than the Phoenix Bird could tell them, so there is no more final message. (Hence, in line with its distributed appearance, one could also interpret this as a message being distributed across space and time and being made of the sum of experiences.)

Another example of distribution is the Borg species from the *Star Trek* universe: This alien species consists of many technologically enhanced individuals (i.e., cyborgs; see Chapter Thirteen) who are all linked among each other and share a common mind and intelligence behind the scenes. All their aspects of individuality are suppressed as they only act *for* and *as* the collective (see Dinello, 2016). In this sense, their physical representation is distributed, while their mind (i.e., their control) is aggregated.

By aggregation we mean that individual parts are combined into a whole in which additional characteristics may emerge. This characterization can be applied to other existing complex systems in life on earth, such as a swarm of bees or an ant colony, a smart factory consisting of many strictly coordinated workers in an assembly line, or a fleet of networked cars which exchange traffic and street conditions among each other, but mainly observe their immediate surroundings. Considering the realm of robots, a solitary social robot could be regarded as a typical example of aggregation, whereas self-organized robots (Pfeifer et al., 2007) with ability such as morphing could be regarded as a more distributed type of robot. One example is a self-reconfigurable robot consisting of many single cubic modules connected through hooks, which enables the robot to change its morphology, e.g., from a quadruped to a linear structure. Still, its single cubicles are perceived as connected and forming a common entity.

The phenomenon where distributed entities consist of several singular smaller entities but are perceived by humans as a unified whole makes use of basic gestalt principles, as described by psychologists in the 1920s. Max Wertheimer, Kurt Koffka, and Wolfgang Kohler wanted to understand how people make sense of the confusing things they see and hear, and so devised a set of principles that address the natural compulsion to find order in disorder. According to their work, "the mind 'informs' what the eye sees by perceiving a series of individual elements as a whole" (Interaction Design Foundation, 2016, para. 4). Perception principles that apply to the above examples are Continuity (grouping elements that seem to follow a continuous path in a particular direction), Common Fate (grouping visual elements moving in the same direction), or Similarity (grouping elements that share superficial characteristics). In sum, such principles ensure that single light dots appear as the Phoenix Bird. Hence, with the Gestalt principles, there is already a good starting point to establish new design paradigms for social robots beyond spatially distinct form. What else could you imagine social robots to be?

If we start from the "Borg" example introduced above, designers could play with the degree of individuality of each member of the collective. On a technical level, we can already assume that future social robots will be networked, so they can actually access and use the shared knowledge of their kind and potentially behave uniformly. However, this might actually not be desired by their human companions, who might appreciate a certain degree of individualism in their robotic companion. Also, it is hard to imagine whether interacting with a collective is something that humans would get used to, or whether it always would feel kind of strange, because one is missing a clear social counterpart.

Naturally, this broader view raises more and more new questions regarding the design, the further we deviate from the familiar interpersonal concepts. If we no longer try to rebuild stereotypes of human assistants and transfer patterns of interpersonal interaction, we have to come up with new interaction paradigms and are confronted with a variety of design decisions: For example, if the robot is distributed is there still a central point of contact? If it does not have a body, can it still express personality? How can it draw the attention of its users, or how do users know it is there and responsive to commands? Is it still adequate to speak of users—or better interaction partners—or attendants (see also von Terzi & Diefenbach, 2023)?

AN EXAMPLE OF A DISTRIBUTED SOCIAL ROBOT: THE ROOM INTELLIGENCE

In the context of a recent research project dealing with the smart home context (PerforM—Personalities for Machinery in Personal Pervasive Smart Spaces), we introduced the concept of the so-called "room intelligence" (RI; see also Diefenbach et al., 2020). In this paradigm, the intelligent interaction partner is not an object in the room, but it essentially *is* the room. Instead of many singular-but-interacting intelligent devices, we propose an overarching interaction concept in which the whole environment (including its parts) is the intelligent (inter)actor. When developing the RI concept, our aim was to provide a comprehensive way of dealing with the increasing complexity of diverse devices assisting us in daily life (e.g., smart heating, smart windows, smart TV, smart fridge, smart coffee machine). In addition to controlling smart devices, the RI is able to manipulate physical objects within the room by mobile robotic arms, and thereby provide support and companionship in daily life. Instead of many separate interfaces and interaction paradigms—one per device or function—the details of operation, movement and analysis are delegated to the RI, thereby conveying the mental model of a central, omnipresent intelligence. While controlling many different devices, the RI represents itself as a coherent entity and social counterpart. If the user wants to activate any of the connected devices, the user enters into a dialogue with the RI, and the RI initiates the necessary operational steps "behind the scenes." Although the effects and points of interaction are distributed across the room, the user perceives this as interacting with a single, coherent entity.

In this sense, the RI can be understood as a distributed form of social robot. This approach is similar to the omnipresent intelligence presented in science fiction movies, such as *Star Trek* or *2001: A Space Odyssey*, (Kubrick, 1968) where the (artificial) intelligence mostly manifests itself as an omnipresent voice. Similarly, the 2009 movie *Moon* (Jones, 2009) shows a physical manifestation as robotic arms acting as "helping hands" and communicating through small displays.

CENTRAL CHARACTERISTICS OF RI

Reflecting on what it needs to establish the envisioned impression of a coherent entity, where the room as a whole is perceived as interaction partner, we came up with several central characteristics, which in a next step can be transformed into concrete technical operationalizations:

- "Invisibility: The RI is acting "behind the scenes." It is not an additional element in the room, but it essentially *is* the room.
- Peripheral interaction: Interaction with the RI is a natural, peripheral action, like talking to someone *en passant*.
- Smooth integration: Blurring the line between the physical and digital worlds, technology becomes an integral part of the environment.
- Cozy atmosphere: The ambience is that of a living room, not a factory floor.
- Transparency: A coherent and understandable mental model for the room's users.
- Coherence: The RI is perceived as one entity, not a collection of singular elements.
- Situational awareness: A sense for the current social situation and adequate actions.
- Personality: The RI can express different personalities.
- Adaptability: The RI shall learn from past interactions and optimize its behavior accordingly" (Diefenbach et al., 2020, p. 2).

INTERACTION CHARACTERISTICS

Regarding the technical operationalization, the room intelligence might combine a number of elements, as we specify elsewhere (see Diefenbach et al., 2020, p. 4). There are "helping hands" which are robotic arms that can move to any part of the room by leveraging a rail system at the ceiling (Fig. 14.1, top) and lend support for specific tasks like cooking; (see Fig. 14.1, second row). There are visual design elements—such as a bright, high-resolution display—that can support interaction. There are also visual elements that offer attractive peripheral stimuli to guide attention; for instance, there is an omnidirectional projective display which can project outputs onto the walls or floor (or really any location), and that has a default "ambient mode," that could turn into "listening mode" or "task-focused mode." In the latter, the light points center around the object involved in the current task such as a coffee machine (see Fig. 14.1, third row). Other elements are acoustic, serving as the "voice" of the room intelligence through distributed speakers. Finally, the room intelligence has a "face" of sorts, accomplished through a schematic representation by fixed or mobile screens or by a grid of mobile cylinders (see Fig. 14.1, bottom).

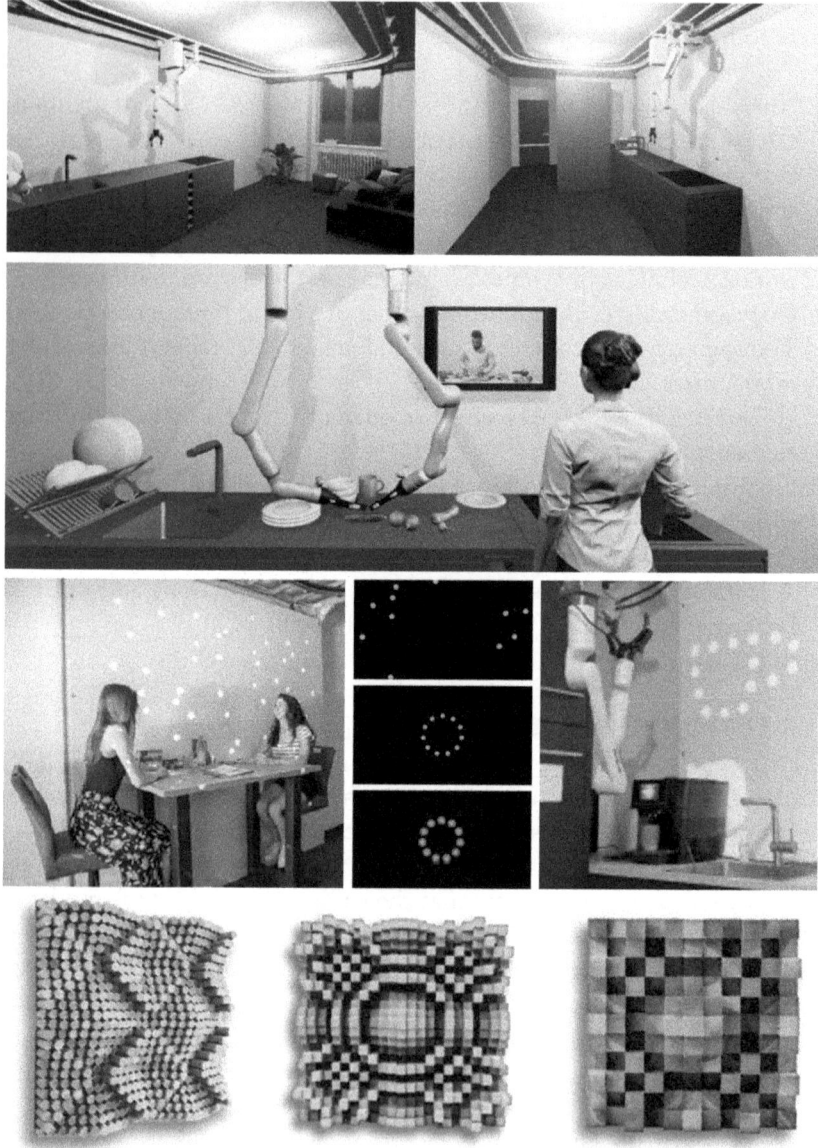

Figure 14.1: Prototypical sketching and experimentation with possible elements of the room intelligence (RI). Top: Robotic arms at a ceiling rail system (source: Authors). Second row: The RI supports specific tasks like cooking (source: Authors). Third row: Guiding attention by light projections indicating different modes such as ambient mode (left), transition from ambient to task mode (middle) and indicating the object of interaction in task mode (right) (source: Daniel Ullrich, in Gräber, 2023). Bottom: The "intraFace," schematic representation of a "face" through a grid of movable cylinders (source: Authors).

QUESTIONS OF PERSONALITY

Finally, besides the constitution of the RI in terms of operational elements, another question is how and to what degree it expresses "personality." Researchers have long emphasized that a technology's outward presentation can influence how humans interact with it and perceive the interaction as more or less satisfying (Nass et al., 1995). As the studies by Nass and colleagues (1995, 2000) showed, people apply rules of social interaction between humans also when interacting with computers or other technical entities. For example, they might literally talk to their smartphone ("Why are you taking so long?"), characterize it with human traits ("Are you stubborn today?"), and (mindlessly) apply norms of courtesy, as if they could hurt the computer's feelings.

Moreover, the conditions under which people perceive computers or robots as trustworthy closely reflect the conditions under which trust becomes fostered in the context of interpersonal communication (Lee & Nass, 2010), whereby the technology's perceived personality traits (e.g., warmth) are one of the influencing components (Christoforakos et al., 2021). As Nass and Moon (2000, p. 93) conclude, "rather than an object for reflection, the computer seems to be a peer in a social interaction." Thus, while the design of a technology and its "personality" crucially impacts how a user perceives and interacts with it, the creation of this impression becomes more complex for distributed entities, where the interplay of several devices shall form the impression of a unified "room personality."

Given the role of an assistant in the home, different types of possible personalities come to mind. For example, the RI could take the role of a caring parent, with a personality that is friendly and helpful but firm and determined when needed. It also could behave more a like a roommate with own needs and suggestions, that requests attention from the user "to care for it." Or it could behave like a professional butler, purely service-oriented, almost eliminating its own personality. Depending on the personality, the RI components could look and act differently. For example, the robotic arms and visual elements could express personality through their speed and form of movement (e.g., stubborn, rapid, relaxed, invariant/monotone) or the intraFace could show different facial expressions.

In the field of human-robot interaction, the question of robot personality and its adequate design is a topic of ongoing debate (for an overview see Diefenbach et al., 2023). However, researchers repeatedly emphasized that robot personality is a vital factor to better understand and facilitate human-robot interaction (e.g., Gockley et al., 2005; Goetz & Kiesler, 2002; Robert, 2018).

Especially for a distributed entity, which is not made visible by a distinct form or body, personality could be an interesting component for a social bond between user and robot. One may even think of "taking my room intelligence" with me to another room, when moving to another place. Still, I recognize it as my RI because it has the same personality—a bit like calling your partner. You recognize it is your partner because of the voice and the way they express themselves, even if you don't see your partner and even if you use another telephone. Independent from the concrete device (the telephone, the room) you are connected with the same social entity (your partner, your RI).

FUTURE VISIONS

The vision of the RI is just one example of what a distributed social robot could look like. Considering the two poles of aggregation and distribution, we must admit that in our daily social interaction, the interaction with an aggregated entity (e.g., other humans, animals) is the more common form, and therefore possibly more intuitive. It is thus reasonable that robot designers often come up with interaction paradigms that are similar to this form of social interaction we are all so familiar with.

If one wants to introduce new paradigms, this has to be done step by step, so that humans can follow—as suggested by the MAYA (most advanced, yet acceptable) principle (e.g., Hekkert, 2006; Loewy, 2002). The MAYA principle promotes a good balance between the status quo and innovation, so that people do not get overwhelmed with too many new concepts or styles beyond their expectations. If introduced step by step, this makes it easier for people to acknowledge the true value of innovations and can lead to success. Thus, though the idea of distributed social actors has not yet entered most peoples' mindset yet this does not mean it does not have potential.

When reflecting on the whole of examples discussed in this chapter, including phenomena from nature (e.g., bee swarms), designed technologies (e.g., room intelligence), or film and fantasy (e.g., borg), it becomes clear that the question of aggregation versus distribution builds a multidimensional design space. Figure 14.2 roughly places some of the discussed examples in a two-dimensional space by their degree of physical distribution and their degree of control distribution.

Thinking about social robots beyond the limits of a distinct body expands our design space to various new possibilities. We could think of single bodies with distributed meta-"minds" (e.g., drone swarms) or about single bodies with some parts distributed across a space rather than attached (in a single room, so still semi-centralized). We hope that this chapter inspires other researchers and designers to think about many other types of distributions which are conceivable, and to have fun with fantasizing about, prototyping, and exploring what else social robots could be.

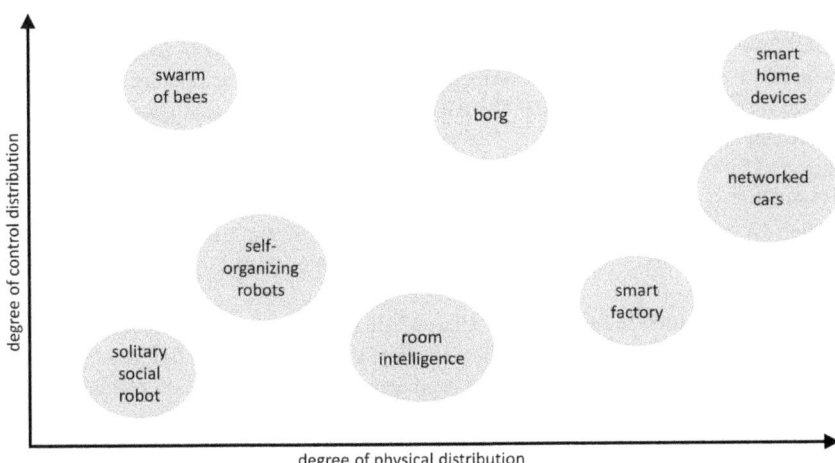

Figure 14.2: Characterizing familiar complex systems by their degree of physical distribution and their degree of control distribution (Source: Authors).

REFERENCES

Bartneck, C., & Forlizzi, J. (2004). A design-centred framework for social human-robot interaction. In *Proceedings of the International Workshop on Robot and Human Interactive Communication* (pp. 591–594). IEEE.

Chen, Y., Wu, F., Shuai, W., & Chen, X. (2017). Robots serve humans in public places—KeJia robot as a shopping assistant. *International Journal of Advanced Robotic Systems, 14*(3).

Christoforakos, L., Gallucci, A., Surmava-Große, T., Ullrich, D., & Diefenbach, S. (2021). Can robots earn our trust the same way humans do? A systematic exploration of competence, warmth and anthropomorphism as determinants of trust development in HRI. *Frontiers in Robotics and AI, 8,* 640444.

Diefenbach, S., Butz, A., & Ullrich, D. (2020). Intelligence comes from within—Personality as a UI paradigm for smart spaces. *Designs, 4*(3), 18.

Diefenbach, S., Herzog, M., Ullrich, D., & Christoforakos, L. (2023). Social robot personality: A review and research agenda. In C. Misselhorn, P. Poljanšek, & T. Störzinger (Eds.), *Emotional machines: Perspectives, affective computing and emotional human-machine interaction* (pp. 217–246). Springer.

Dinello, D. (2016). The Borg as contagious collectivist techno-totalitarian transhumanists. In K. S. Decker & J. T. Eberl (Eds.), *The ultimate Star Trek and Philosophy: The search for Socrates* (pp. 83–94). Wiley.

Gräber, L. (2023). *Enhancing trust in human-robot interaction by using visual cues to guide and communicate attention* [Unpublished Master's Thesis, Ludwig-Maximilians-Universität München].

Gockley, R., Bruce, A., Forlizzi, J., Michalowski, M., Mundell, A., Rosenthal, S., ... & Wang, J. (2005). Designing robots for long-term social interaction. In *Proceedings of the International Conference on Intelligent Robots and Systems* (pp. 2199–2204). IEEE/RSJ.

Goetz, J., & Kiesler, S. (2002). Cooperation with a robotic assistant. In *Extended Abstracts of the CHI Conference on Human Factors in Computing Systems* (pp. 578–579). ACM.

Heerink, M., Kröse, B., Evers, V., & Wilinga, B. (2010). Assessing acceptance of assistive social agent technology by older adults: The Almere Model. *International Journal of Social Robotics, 2*, 361–375.

Hekkert, P. (2006). Design aesthetics: Principles of pleasure in design. *Psychology Science, 48*(2), 157–172.

Interaction Design Foundation—IxDF. (2016). *What are the Gestalt principles?* Interaction Design Foundation—IxDF. <https://www.interaction-design.org/literature/topics/gestalt-principles>

Jones, D. (Director). (2009). *Moon* [Film]. Stage 6 Films; Liberty Films; Xingu Films; Limelight.

Kubrick, S. (Director). (1968). *2001: A Space Odyssey* [Film]. Stanley Kubrick Productions.

Lee, J. E. R., & Nass, C. I. (2010). Trust in computers: The computers-are-social-actors (CASA) paradigm and trustworthiness perception in human-computer communication. In *Trust and technology in a ubiquitous modern environment: Theoretical and methodological perspectives* (pp. 1–15). IGI Global.

Loewy, R. (2002). *Never leave well enough alone.* JHU Press.

Nass, C., Moon, Y., Fogg, B. J., Reeves, B., & Dryer, C. (1995). Can computer personalities be human personalities? *International Journal of Human-Computer Studies, 43*(2), 223–239.

Nass, C., & Moon, Y. (2000). Machines and mindlessness: Social responses to computers. *Journal of Social Issues, 56*(1), 81–103.

Pfeifer, R., Lungarella, M., & Iida, F. (2007). Self-organization, embodiment, and biologically inspired robotics. *Science, 318*(5853), 1088–1093.

Robert, L. (2018). *Personality in the human robot interaction literature: A review and brief critique.* In *Proceedings of the Americas Conference on Information Systems* (pp. 1–10). SSRN.

Stone, J. (Director). (1983). *Big Bird in China* [Film]. Children's Television Workshop.

Von Terzi, P., & Diefenbach, S. (2023). The attendant perspective: Present others in public technology interactions. In *Proceedings of the CHI Conference on Human Factors in Computing Systems* (no. 502, pp. 1–18). ACM.

Wachowski, L., & Wachowski, L. (Directors). (2003). *The Matrix Revolutions* [Film]. Warner Bros. Pictures; Village Roadshow Pictures; NPV Entertainment; Silver Pictures.

PART II
Implicit Anatomy

CHAPTER FIFTEEN

Images & Frames: Tensions in Representations

AIKE C. HORSTMANN

Social robots are specifically designed to adopt the role of a social interaction partner (Dautenhahn, 2007; Wiese et al., 2017) that triggers conscious and unconscious social reactions (Guthrie, 1997; Reeves & Nass, 1996; Seibt et al., 2020). Since social robots are of technological nature but act and appear somewhat alive (see Chapter Twenty Four), they are difficult to compare to other technologies and thus hard to classify (Kahn et al., 2011). However, humans have a pronounced need to classify and structure their (social) environment in order to reduce uncertainty and make reliable predictions (Berger & Calabrese, 1975). Since most people have not personally interacted with a social robot yet, they need to draw on sources other than their own experiences for information. Here, mass media portrayals (visual imagery as well as textual representations) are assumed to shape people's images of social robots tremendously (Banks, 2020; Bruckenberger et al., 2013; Horstmann & Krämer, 2019). As a result, people may rely on these images as media-informed *frames* which affect how they expect, perceive, and interpret social robots' behaviors. Frames are cognitive filters that help people to process and structure their immediate experience (Goffman, 1974). Ultimately, frames serve as lenses to organize and interpret incoming information and to guide people in how to approach situations that are novel but are similar to previous situations (Banks et al., 2021; Scheufele, 2000). Therefore, when people approach a social robot, how they perceive and interpret the robot and its behavior may be heavily affected by media-informed frames.

IMAGES OF ROBOTS IN MASS MEDIA

Social robots are frequently portrayed in mass media: They often play a major role in diverse science fiction formats, which range from movies and games to comics and books, but can also be the subject of reports, documentaries, and news articles. In general, mass media represent social robots as agentic entities (Banks, 2020) with advanced technical, cognitive, and physical abilities that enable them to navigate complex social settings and perform a variety of activities that are usually performed by humans (Bruckenberger et al., 2013; Kriz et al., 2010; Oliveira & Yadollahi, 2023; Sandoval et al., 2014). For instance, the titular character in *Mrs. Doubtfire* can move about as a human does: Doing chores, telling stories, solving problems, and having complex conversations. In fictional media, social robots are even portrayed to navigate socially complex, intimate human–robot relationships as sentient beings with the ability to experience feelings and exert free will (Döring & Poeschl, 2019).

Besides presenting social robots with advanced and human-like skills, science fiction tends to portray them in a polarized way as either "good" (e.g., superheroes trying to save the planet or very helpful assistants) or "bad" (e.g., evil, intelligent robot villains trying to overturn and suppress humans). Since either the robots' prosocial abilities or their violent, destructive motives are emphasized, the relationship between humans and robots tends to be either portrayed as friendship or antagonism (Oliveira & Yadollahi, 2023). These polarized portrayals create a mixed perception of real robots, resulting in conflicting emotions toward them (Bruckenberger et al., 2013).

On the one hand, common themes in science fiction include the dangerous machine, robots desiring life or consciousness, and the overly intelligent robot (Khan, 1998). These portrayals can foster negative images of robots. For instance, a widespread fear is that robots will develop their own agenda and revolt against humans (Khan, 1998; Nomura et al., 2012) which is reflected in the storylines of movies such as *The Terminator*, *The Matrix*, and *I, Robot*. Commonly depicted concerns that robots will compete with, replace, and dominate humans are referred to as Frankenstein syndrome or complex (Asimov, 1947; Nomura et al., 2012). This is further reflected by the Three Laws of Robotics presented by the popular science fiction author Asimov (1947, 1950) which are frequently mentioned by the media in the context of human-robot interaction (Murphy & Woods, 2009). Asimov's laws (minimizing harm to humans, obeying human orders, and self-preservation) were originally developed as a literary device for entertaining storytelling but have since shaped society's expectations about how robots should act around humans and have been taken up as frameworks for robot design, policymaking, and ethical considerations (Murphy & Woods, 2009).

On the other hand, recent analyses found media to portray robots as friendly, helpful companions with pronounced social abilities, rather than harmful competitors (Dieter & Gessler, 2021; Oliveira & Yadollahi, 2023). Particularly, modern films often depict cohesive and complimentary interactions between social robots and humans (Dieter & Gessler, 2021). For instance, *Frank & Robot* describes the development of a friendship between a retiree and his home-care robot and *Big Hero 6* tells the story of an oversized, inflatable robot that takes care of a teenager, who in turn optimizes him technically and aesthetically. These depictions may be intended to illustrate scenes of a preferred reality where robots are meant to be our helpers rather than threatening competitors (Dieter & Gessler, 2021).

It has been argued that in Western cultures there is a stronger focus on the dangerous and evil robot, while in Eastern cultures robots are depicted as good and as fighting against other evil entities (Bartneck et al., 2005). However, Western pop culture also offers various good-willed robots such as *WALL-E*, *Bumblebee* from *The Transformers*, and *C-3PO* from *Star Wars*. Western storytelling typically depicts a main character pursuing a specific goal and villains who want to stop the protagonist: A fight between clearly good and evil, where good should win (Shah et al., 2023). Eastern storytelling is usually more complex: Writers create characters whose main drive is to do something good for society and if there are antagonists, they are initially good and struggling because they were manipulated or have been lied to (Shah et al., 2023). One of the main differences may therefore be that the real fight is not cut and dry (i.e., good vs. evil) but to find the good that is in everything and to help it thrive. Across these differences in depiction, representations of robots broadly inform the frames people engage to consider actual robots.

FRAMES TO INTERPRET SOCIAL ROBOTS' BEHAVIORS

When encountering a new interaction partner, humans strive to reduce the uncertainty about the other as soon as and as much as possible (Berger & Calabrese, 1975). The aim is to be able to understand and predict the other's behavior, and people draw from past experiences and other available sources of information to do so. In a similar sense, Goffman (1974) postulates that individuals generally classify and interpret their life experiences to understand the world around them which results in schemata of interpretation called frameworks or frames. These frames can be seen as filters that guide expectations and influence the perception and interpretation of an event. More specifically, frames ascribe relevance and consequently attention to certain aspects while others, which might have been at the center of focus with a different frame, are neglected or ignored (Nelson et al., 1997).

There are individual frames which are defined as native understandings or "mentally stored clusters of ideas that guide individuals' processing of information" (Entman, 1993, p. 53) and produced frames (or media frames) which are discursive structures or characterizations by the media or other third parties (Banks & Koban, 2021; Scheufele, 1999). Individual and produced frames are not necessarily distinct from each other, for instance information from the media (produced frames) can contribute to people's knowledge basis to form individual frames (Banks & Koban, 2021).

When people encounter a social robot, individual frames and produced frames come into play. For people who have not (yet) personally interacted with a social robot, representations of robots in mass media may primarily inform their individual frames (Banks, 2020; Kriz et al., 2010; Weiss et al., 2011). Mass media are widely accessible and therefore a significant first source for individuals to obtain information about a specific technology when they do not have personal access to it (Kriz et al., 2010; Sundar et al., 2016). Science fiction representations are particularly effective since they incorporate a narrative structure. Here, research found that introducing a social robot and its features with a narrative story, in contrast to an instruction manual, leads to a more positive perception of the robot (e.g., more useful, likable, intelligent, autonomous, and humanlike) and a higher willingness to continue interacting with it (Mara et al., 2013; Rosenthal-von der Putten et al., 2017). The representations of social robots in popular culture have the potential to inform people's interpretative frames which ultimately shapes their perceptions of and attitudes towards them (Khan, 1998; Weiss et al., 2011). As a result, mass media may cultivate a shared understanding in societal groups of what social robots are and how they act (Banks, 2020; Gerbner & Gross, 1976).

SOCIAL AND PSYCHOLOGICAL EFFECTS OF IMAGES AND FRAMES

People are worried about autonomous robots, humans being replaced by robots, loss of control over as well as dysfunction of robots (Ray et al., 2008)—fears that match many typical science fiction scenarios. In line with that, research showed that the ability to recall more "bad" robot characters from science fiction, such as the *T-800* from *The Terminator* or the *NS-5 robots* from *I, Robot*, the more negative expectancies of robots becoming a threat to humans they have (Horstmann & Krämer, 2019). A greater recollection of robot characters that are human-like and/or elicit sympathy, such as the *Bicentennial Man* or *Sunny* from *I, Robot*, appears to be linked to lower anxiety toward robots (Sundar et al., 2016). When people watched both "bad" and "good" media representations of robots, they develop emotional reactions perhaps without fully realizing it: The more sympathy individuals have for robot characters recalled from media, the more likely they are to engage in future collaborations with an actual robot, to trust it, and to perceive it has having a mind (Banks, 2020).

Media portrayals may further lead to biased expectations of robots' capabilities in the real world since they are often depicted as having advanced cognitive and physical abilities (Bruckenberger et al., 2013; Kriz et al., 2010; Sandoval et al., 2014). It was found that the reception of social robots in the media (particularly science fiction) leads to higher expectations regarding their level of skills and how much they will be part of the society and people's personal lives in the future (Horstmann & Krämer, 2019). However, the commercially available robots we have today are far behind fictional robots in terms of both physical and cognitive abilities (Broadbent, 2017; Oliveira & Yadollahi, 2023). Even when a fictional robots' appearance mirrors that of a real-life robot, such as the robot from *Robot & Frank* which resembles the ASIMO robot developed by Honda, the fictional robot's abilities far exceed the abilities of the real one (Broadbent, 2017). A movie with an accurate representation of the current state of real-life robots would likely not attract much attention. In a similar way, news articles may also put a focus on what sells the story (Oliveira & Yadollahi, 2023). As a result, media reports are characterized by an overemphasis on successes and progresses of social robots and an underrepresentation of shortcomings, unsolved technological problems, and setbacks (Weiss et al., 2011). As Bartneck and colleagues (2020) argue, a news article titled "Robots are harmless and almost useless" is unlikely to attract attention, although that would be a fitting description of most real-life social robots at the present time.

When people use frames informed by mass media, there will likely be mismatches and expectancy violations that may negatively affect how social robots and their behaviors are interpreted. Most concerns and fears regarding robots are based on a media-induced overestimation of their capabilities. But also, individuals who are excited about what robots are capable of—based on how they are depicted in science fiction and in reports of actual robots—will be highly disappointed when they encounter real robots (Horstmann & Krämer, 2019). These expectancy violations will cause more negative feelings towards social robots than when they would have been portrayed in an accurate, but potentially less exciting way from the beginning. Media misrepresentations may even have far-reaching consequences. For instance, they can lead to a misdirection of public debate about important topics related to technology and the passing of regulation that is disconnected from the reality in terms of scientific and technological development (Cave et al. 2018).

However, media-produced frames can also be resisted, for instance when they do not align with individual frames (Scheufele, 1999). Particularly, a single produced frame may generally not be able to overwrite pre-existing interpretative lenses. For instance, people with a high technophobia were observed to reject a frame produced by a robot and to hold on to their individual frame instead (Banks & Koban, 2021). In human-robot interaction, people were particularly observed to neglect the information from produced frames after an actual interaction with a social robot.

IMPLICATIONS FOR THE REPRESENTATION AND DESIGN OF SOCIAL ROBOTS

The previous sections lay out how social robots are imaged in mass media, how these images help to inform frames, and how these frames shape individual's interpretation and so their responses to social robots. One implication would be that if media should portray social robots more realistically, audiences would be able to develop more accurate expectations before they actually interact with the technology. In doing so, first encounters might not be anxiety-producing or disappointing. However, it needs to be considered that science fiction has an important role of conceptually prototyping future technologies which extends and enhances the traditional practices of research and design (Johnson, 2011). For instance, the invention of submarines was inspired by Jules Verne's *Twenty Thousand Leagues Under the Sea*, the design of mobile phones was shaped by Captain Kirk's communicator in *Star Trek*, and long before Apple presented its Siri software the robot *R2-D2* from *Star Wars* was controlled via voice command. By exploring potential use cases and respective functionalities of social robots, science fiction can help to question and redefine current thinking and to shape the meanings held for novel technologies (Appel et al., 2016). Indeed, these fictions and high-science innovations are known to shape each other (Fleischmann & Templeton, 2008). The next generation of science fiction should leave behind the question of whether robots will be humans' enemies or allies and instead focus on prototyping our future life with social robots—what it may mean to live, work, play, and even feel along with them. As science fiction can break through current conceptual barriers it holds the potential to inspire and spark technological progress in ways that were not thought about before.

REFERENCES

Appel, M., Krause, S., Gleich, U., & Mara, M. (2016). Meaning through fiction: Science fiction and innovative technologies. *Psychology of Aesthetics, Creativity, and the Arts, 10*(4), 472–480.
Asimov, I. (1947). *Little lost robot.* Street & Smith.

Asimov, I. (1950). *I, robot*. Gnome Press.
Banks, J. (2020). Optimus primed: Media cultivation of robot mental models and social judgments. *Frontiers in Robotics and AI, 7*, 62.
Banks, J., & Koban, K. (2021). Framing effects on judgments of social robots' (im)moral behaviors. *Frontiers in Robotics and AI, 8*, 627233.
Banks, J., Koban, K., & Chauveau, P. (2021). Forms and frames: Mind, morality, and trust in robots across prototypical interactions. *Human-Machine Communication, 2*, 81–103.
Bartneck, C., Belpaeme, T., Eyssel, F., Kanda, T., Keijsers, M., & Šabanović, S. (2020). *Human-robot interaction: An introduction*. Cambridge University Press.
Bartneck, C., Nomura, T., Kanda, T., Suzuki, T., & Kennsuke, K. (2005). A cross-cultural study on attitudes towards robots. In *Proceedings of the International Conference on Human-Computer Interaction* (no. 25). MIRA.
Berger, C. R., & Calabrese, R. J. (1975). Some explorations in initial interaction and beyond: Toward a developmental theory of interpersonal communication. *Human Communication Research, 1*(2), 99–112.
Broadbent, E. (2017). Interactions with robots: The truths we reveal about ourselves. *Annual Review of Psychology, 68*, 627–652.
Bruckenberger, U., Weiss, A., Mirnig, N., Strasser, E., Stadler, S., & Tscheligi, M. (2013). The good, the bad, the weird: Audience evaluation of a "real" robot in relation to science fiction and mass media. In *Proceedings of the International Conference on Social Robotics* (pp. 301–310). Springer.
Cave, S., Craig, C., Dihal, K., Dillon, S., Montgomery, J., Singler, B., & Taylor, L. (2018). *Portrayals and perceptions of AI and why they matter*. The Royal Society.
Dautenhahn, K. (2007). Socially intelligent robots: Dimensions of human-robot interaction. *Philosophical Transactions of the Royal Society of London. Series B, Biological Sciences, 362*(1480), 679–704.
Dieter, D. G., & Gessler, E. C. (2021). A preferred reality: Film portrayals of robots and AI in popular science fiction. *Journal of Science & Popular Culture, 4*(1), 59–76.
Döring, N., & Poeschl, S. (2019). Love and sex with robots: A content analysis of media representations. *International Journal of Social Robotics, 11*(4), 665–677.
Entman, R. M. (1993). Framing: Toward clarification of a fractured paradigm. *The Journal of Communication, 43*(4), 51–58.
Fleischmann, K. R., & Templeton, T. C. (2018). Past futures and technoscientific innovation: The mutual shaping of science fiction and science fact. *Proceedings of the American Society for Information Science and Technology, 45*(1), 1–11.
Gerbner, G., & Gross, L. (1976). Living with television: The violence profile. *The Journal of Communication, 26*(2), 173–199.
Goffman, E. (1974). *Frame analysis: An essay on the organization of experience*. Harvard University Press.
Guthrie, S. E. (1997). Anthropomorphism: A definition and a theory. In R. W. Mitchell, N. S. Thompson, & H. L. Miles (Eds.), *Anthropomorphism, Anecdotes, and Animals* (pp. 50–58). State University of New York Press.
Horstmann, A. C., & Krämer, N. C. (2019). Great expectations? Relation of previous experiences with social robots in real life or in the media and expectancies based on qualitative and quantitative assessment. *Frontiers in Psychology, 10*, 939.

Johnson, B. D. (2011). *Science fiction prototyping: Designing the future with science fiction*. Morgan & Claypool Publishers.

Kahn, P. H., Reichert, A. L., Gary, H. E., Kanda, T., Ishiguro, H., Shen, S., ... & Gill, B. (2011). The new ontological category hypothesis in human-robot interaction. In A. Billard, P. Kahn, J. A. Adams, & G. Trafton (Eds.), *Proceedings of the International Conference on Human-Robot Interaction* (pp. 159–160). ACM.

Khan, Z. (1998). *Attitudes towards intelligent service robots*. Royal Institute of Technology.

Kriz, S., Ferro, T. D., Damera, P., & Porter, J. R. (2010). Fictional robots as a data source in HRI research: Exploring the link between science fiction and interactional expectations. In *Proceedings of the International Symposium in Robot and Human Interactive Communication* (pp. 458–463). IEEE.

Mara, M., Appel, M., Ogawa, H., Lindinger, C., Ogawa, E., Ishiguro, H., & Ogawa, K. (2013). Tell me your story, robot. Introducing an android as fiction character leads to higher perceived usefulness and adoption intention. In *Proceedings of the International Conference on Human-Robot Interaction* (pp. 193–194). IEEE.

Murphy, R., & Woods, D. D. (2009). Beyond Asimov: The three laws of responsible robotics. *IEEE Intelligent Systems, 24*(4), 14–20.

Nelson, T. E., Clawson, R. A., & Oxley, Z. M. (1997). Media framing of a civil liberties conflict and its effect on tolerance. *American Political Science Review, 91*(3), 567–583.

Nomura, T., Sugimoto, K., Syrdal, D. S., & Dautenhahn, K. (2012). Social acceptance of humanoid robots in Japan: A survey for development of the Frankenstein Syndrome Questionnaire. In *Proceedings of the International Conference on Humanoid Robots* (pp. 242–247). IEEE.

Oliveira, R., & Yadollahi, E. (2023). Robots in movies: A content analysis of the portrayal of fictional social robots. *Behaviour & Information Technology, 43*(5), 970–987.

Ray, C., Mondada, F., & Siegwart, R. (2008). What do people expect from robots? In *Proceedings of the International Conference on Intelligent Robots and Systems* (pp. 3816–3821). IEEE.

Reeves, B., & Nass, C. (1996). *The media equation: How people treat computers, television, and new media like real people and places*. Cambridge University Press.

Rosenthal-von der Pütten, A., Strassmann, C., & Mara, M. (2017). A long time ago in a galaxy far, far away...The effects of narration and appearance on the perception of robots. In *Proceedings of the International Symposium on Robot and Human Interactive Communication* (pp. 1169–1174). IEEE.

Sandoval, E. B., Mubin, O., & Obaid, M. (2014). Human robot interaction and fiction: A contradiction. In M. Beetz, B. Johnston, & M.-A. Williams (Eds.), *Lecture Notes in Computer Science. Social Robotics* (Vol. 8755, pp. 54–63). Springer.

Scheufele, D. A. (1999). Framing as a theory of media effects. *The Journal of Communication, 49*(1), 103–122.

Scheufele, D. A. (2000). Agenda-setting, priming, and framing revisited: Another look at cognitive effects of political communication. *Mass Communication and Society, 3*(2–3), 297–316.

Seibt, J., Vestergaard, C., & Damholdt, M. F. (2020). Sociomorphing, not anthropomorphizing: Towards a typology of experienced sociality. In M. Nørskov, J. Seibt, & O. S. Quick (Eds.), *Frontiers in Artificial Intelligence and Applications. Culturally Sustainable Social Robotics* (pp. 51–67). IOS Press.

Shah, M. R., Ahmad Rafi, M. E., & Perumal, V. (2023). Investigating storytelling differences between western and eastern computer animation. In F. Mustaffa, R. Sitharan, & J. S. Mohd

Nasir (Eds.), *Proceedings of the International Conference on Creative Multimedia* (pp. 124–133). Atlantis Press SARL.

Sundar, S. S., Waddell, T. F., & Jung, E. H. (2016). The Hollywood Robot Syndrome: Media effects on older adults' attitudes toward robots and adoption intentions. In *Proceedings of the International Conference on Human-Robot Interaction* (pp. 343–350). IEEE.

Weiss, A., Igelsböck, J., Wurhofer, D., & Tscheligi, M. (2011). Looking forward to a "robotic society"? *International Journal of Social Robotics, 3*(2), 111–123.

Wiese, E., Metta, G., & Wykowska, A. (2017). Robots as intentional agents: Using neuroscientific methods to make robots appear more social. *Frontiers in Psychology, 8*, 1663.

CHAPTER SIXTEEN

Digitality & Interactivity: Lessons Learned from NPCs

NICHOLAS DAVID BOWMAN, ELENA YIFEI ZHAO, &
YOON ESTHER LEE

"It's dangerous to go alone! Take this."
~ unnamed man, from *The Legend of Zelda* (Nintendo, 1986)

"What is a man? A miserable little pile of secrets."
~ Dracula, from *Castlevania: Symphony of the Night* (Konami, 1997)

"We are pleased that you made it through the final challenge, where we pretended we were going to murder you."
~ GLaDOS, from *Portal* (Valve, 2007)

The opening quotes for this chapter are pulled from human-machine dialogues—in each case, human video game players are engaging with non-player characters, or NPCs. Although not all video games are populated with NPCs, many of them are, and NPCs such as the ones above are the social agents occupying and making more authentic their respective digital worlds (i.e., video games).

Although not formally defined as social robots (and indeed, our goal is not to define NPCs as social robots *per se*), NPCs exemplify the social apparatus key to definitions of social robots (Hegel et al., 2009). NPCs are "robots" in the sense that they are automated agents, constructed of digital code embodied in pixels and polygons, executing their respective scripts as programmed. They are "social" in the sense that they interact with players adventuring in the respective digital worlds and provide the backdrop for much of the social interaction that takes place in these games. The more compelling and engaging NPCs are presented as if they have sentience, offering players a convincing intelligence to interact with (i.e., affective computing; Hamdy & King, 2017).

In these opening notes, two terms have emerged that are equally critical for NPCs and social robots: Digitality and interactivity. By digitality, we are referring to the binary code (Heath, 1972) that results in the on-screen content displayed to the users. By interactivity, we refer to the user's ability to variable alter and modify the form and content of those displays (*interacting with the content*; Steuer, 1992), although we could also refer to conversational give-and-take (*interacting with other agents*; Rafaeli & Sudweeks, 1997) between social agents in the environment, or even considering games as an ongoing human-computer interaction (Cardona-Rivera & Young, 2014; Bowman, 2016).

In this chapter, we introduce canonical media and communication theories and perspectives consistently applied to gaming experiences—social presence and electronic propinquity, player-avatar relations and interactions, and interactivity-as-demand models. For each of these theories and perspectives, we elaborate on how that scholarship has been used to explain NPCs in gaming environments, and then extend those discussions to how the same scholarship can help us better understand how humans engage with social robots.

SOCIAL PRESENCE AND ELECTRONIC PROPINQUITY

Originating from the work of Short et al. (1976), social presence is a phenomenon in which users engage mediated-communication environments—environments that *de facto* separate users at some distance—yet those users feel a sense of *being with* others while inside that digital world (see Lombard & Ditton, 1997). Critically, social presence is not a product of the medium but rather of the users' perceptual and psychological engagement with the medium (Lee, 2004). A related yet lesser-known concept related to social presence is electronic propinquity, understood by Korzenny (1978) as feeling a spatial or temporal nearness to another during online interactions—a psychological closeness experienced by those engaged in communication (Walther & Bazarova, 2008). Both concepts have had a substantial impact on how we understand the experiences of users with online avatars (Banks & Carr, 2019), social robots (Hoorn, 2020), and other internet-of-things technologies (Kang & Kim, 2020). That said, the concepts have a key distinction: Social presence does *not* require one to interact with another, while electronic propinquity is a *de facto* feature of agent-to-agent communication. That said, both are useful concepts for understanding how we perceive social others, even if those "others" are not humans.

Application to Video Games and Digital Worlds

Many video games are designed to cultivate social experiences through features such as multiplayer modes and collaborative gameplay (Ekman et al., 2012)—thus helping players feel with, among, and close to other humans in the digital space. The inclusion of visible characters within the gaming environment promotes social interactions among players (Tamborini & Bowman, 2010). Embodied avatars foster higher levels of social presence, and these effects held regardless of how cartoony or realistic the avatars were (Liszio et al., 2017; Yoon et al., 2019). In various gaming contexts, myriad types of character styles result in different avatar types being used (see Junuzovic et al., 2012; Jo et al., 2017)—from humans and orcs in *Warcraft* to racecars in *Burnout: Paradise*.

Social presence and propinquity are essential to engaging with non-player characters (NPCs) and digital agents broadly. When NPCs are seen as autonomous, players feel a heightened sense of responsibility for and attachment to those digital agents, viewing them as challenging-yet-engaging companions (Brandstätter et al., 2021). The heightened social presence derived from NPCs in virtual reality (VR) based video games is correlated with reductions in perceived loneliness (Liszio et al., 2017) and the elicitation of more positive emotional responses to the experience (Sajjadi et al., 2019). Players can also engage in forced perspective-taking with NPCs—literally, seeing digital worlds through the eyes of the other—and when they do, report increased psychological closeness (similar to electronic propinquity) with and empathy towards the NPCs, as well as increased game immersion as a whole (Ho & Ng, 2022). Likewise, Schumann et al. (2016) found that positive interactions with NPCs throughout the game led to presence; feeling as if the player is inside the game environment (also see Holl & Melzer, 2022).

Application to Social Robots

Social presence and electronic propinquity are useful concepts as they underlie our perceptions of social others. Edwards et al. (2019) found social presence to increase (and uncertainty to decrease) after an initial encounter with a humanoid social robot. Pereira et al. (2014) discovered that when humans felt social presence with social robots during gameplay, they felt as if the robot had much higher social skills. Yi-Chen Chen et al. (2023) found that social presence motivated children to engage with social robots, whereas increased and ongoing interactions (similar to propinquity) help foster rapport, and Montalvo et al. (2022) found that individuals higher on trait loneliness felt increased social presence with a social robot. Na Chen et al. (2023) have even developed a *robot social presence* scale to further capture these feelings. One possibility here is that facilitating social presence with social robots is likely to encourage a sense of electronic propinquity, driving future human-robot interactions. Such a potential is already seen in Haggadone et al. (2021), who found that spending time with an otherwise-intimidating large humanoid robot resulted in lowered perceived social distance between participants and the robot itself.[1]

PLAYER-AVATAR RELATIONS, INTERACTIONS

Focusing a bit more on (social) interactions with digital agents, Banks (2015) offers a typology that more deeply explains the ways in which we engage avatars in games and digital worlds. The player-avatar relations model (PAR) argues that when we engage with digital avatars, we do so somewhere along a continuum of sociality. At the lowest levels of sociality (i.e., asocial), players see avatars as objects—viewing them as tools or toys or similar metaphors (see Banks & Bowman, 2015). From this, players might see the avatar as "me"—seeing themselves reflected in the on-screen avatar. That said, players need not psychologically merge with these on-screen agents, as social interactions require multiple, independent entities. Thus, some players engage a more symbiotic relation with their avatar—seeing bits of themselves but also bits of an (often, idealized) other social agent. At the far end of this sociality continuum is a fully realized social relationship in which the player sees the avatar as a separate social agent, with its own motivations and perspective. Although not studied prior, one could presume that Avatar-as-Other orientations would likely correlate with increased social presence and electronic propinquity with on-screen avatars. Underlying these PAR categorizations are more discrete relational dynamics, captured using the player-avatar interaction scale (PAX; Banks & Bowman, 2016; Banks et al., 2019). PAX dimensions include relational closeness (feeling an emotional attachment to and interdependence with the avatar), anthropomorphic autonomy (seeing the avatar as having a human-like agency distinct from the player), critical concern (attending to the coherence of the avatar's existence in its digital world), and a sense of control (feeing functional governance over the avatar's actions).

---

[1] Reports of the robot's intimidating nature were shared by the researchers of the study, who remarked that several participants refused to take part after being seated in front of their lab's Robothespian (Jaime Banks, personal communication, 8 February 2024).

Application to Video Games and Digital Worlds

While social presence and electronic propinquity emerged as a way to understand relational closeness in computer-mediated relations (usually without an embodied "other" during the interaction), PAR and PAX were borne from scholarship in gaming and digital worlds and thus organize our understanding of how players engage and interact with digital agents in those spaces. For example, object orientations led to the lowest feelings of identification with avatars, and increased liking for and perspective-taking with the avatar were correlated with other PAR orientations (Bowman, Banks, & Downs, 2021). That same study found that feelings of embodiment, value homophily, and physical similarity were highest for me orientations (suggesting a level of player-avatar identification). Banks and Bowman (2016b) found increased use of third-person singular pronouns to describe avatars seen as symbiotes and others (others being more prevalent) and likewise a general absence of pronouns used to describe avatars seen as objects.

Application to Social Robots

Just as player interactions with digital NPCs can be mapped to how we might engage with social robots, we also draw comparisons between how players teleoperate their avatars in on-screen digital worlds, and how they might similarly operate and interact with others through robots. Indeed, the remote controlling of avatars is increasingly prevalent in social-robot control schemes, such as that for the robot Reachy (Pollen Robotics, n.d.) that is controlled using a VR headset and handheld controls. Minsky (1980) referred to *telepresence* to explain feelings of non-mediation during these interactions such that users felt as if they were directly in the remote environment, and Topal et al. (2010) suggested that such teleoperation could be understood as an "avatarization"—human agency inserted in the remote location to facilitate interactions at a distance.[2] Establishing this connection was a critical component of Bowman and Banks' (2024) arguments that the PAR and PAX approaches could be directly applied to how humans might variably relate to their own (variably social) robots. Similar to avatars in digital worlds, people likely range considerably in how they engage and interact with their own robots: Object orientations might be most likely for task-focused relations (such as those in which a human routinely uses a robot to move and manipulate objects in factory) whereas me orientations might be most likely for

---

[2] Curious readers will note the conceptual similarity between Minsky's notions of telepresence and our earlier discussions of social presence. Both Steuer (1992) and Lombard and Ditton (1997) offer an extensive summary of the similarities between these constructs and on the centrality of Minsky's claims to later work on presence more broadly.

highly social relations in which the user's own identity is important (such as those in which the robot is routinely engaged to interact with co-workers in collaborative work). Of course, just as gamers need not *de facto* see themselves in their on-screen avatars, teleoperated robots need not serve as mere vessels for human identities. For example, Häkkilä et al. (2022) found that some users expressed a desire to craft social robots in ways that *diminished* their similarity to the human teleoperator, perhaps to allow the robots to have a unique identity. Such arguments are central to where and how we might see robots as having moral patiency (Banks et al., 2021; Gunkel & Wales, 2021), which is the consideration for their well-being that is key to forming human-like relationships.

THE MANY DEMANDS OF INTERACTIVITY

Finally, we can discuss a more fundamental approach to how we understand interactivity—or at least, what it means to engage an interactive system. For example, in video games and digital worlds, part of these experiences is that players are cast as co-authors of the experiences (Wellenreiter, 2015). By this, we mean that players make choices in response to on-screen content and in this way, they directly influence the experience in variably idiosyncratic ways. Bowman (2021) argued that interactivity is a near-constant demand on player's resources, including cognitive demands (a requirement to rationalize and understand challenges), emotional demand (explicit or implicit affective responses to in-game events), physical demands (exerting discrete or holistic efforts into manipulating game controls), and social demands (responses to other social actors, both human and non-human).[3]

Application to Video Games and Digital Worlds

Key to the interactivity-as-demand model is a claim that *demands are unlikely to operate at simultaneously high levels*. Indeed, game designers recognize potentially competing demands. For example, using "story mode difficulty" (see TVTropes.org, n.d.) to allow players to focus on the emotional and social demands of narratives and characters within a digital storyworld also means lowering the cognitive and physical demands required of otherwise intense challenges. Notably, this same example shows where demands might complement each other, and this claim has

---

3 Although a comparatively newer concept, the dimensionality of the interactivity-as-demand model has been replicated in several studies, including measurement translations to English, German, Korean, Mandarin Chinese, Portuguese, Spanish, and Turkish (linked via open science documentation at https://osf.io/x5jch/).

been corroborated in prior research. Lin et al. (2022) found that playing video games with a virtual reality headset increases the perceived cognitive and physical demands of the experience—the former having a positive impact on enjoyment and the latter having a marginally negative impact on the same. Bowman, Keene, and Najera (2021) reported that when the challenge-skill balance was manipulated in video games from boring (low challenge and high skill) to frustrating (high challenge and low skill), players felt that the games were increasingly cognitively and physically demanding. Demands have also been studied in non-gaming interactive media in ways that explained suboptimal user engagement and other outcomes. These include increased cognitive demands associated with using virtual reality for practicing speeches (possibly hindering learning outcomes; Kryston et al., 2021) and increased physical demands associated with using VR headsets to view videos of natural disasters (hindering empathy for those affected; Pressgrove & Bowman, 2021). Others have studied demand imbalances as a contributor to observed trolling behaviors (Cook, 2019) as well as where demands might help in creating video game experiences aimed at therapeutic use (Frommel et al., 2021), and the model has been integrated into broader theories of game-based learning (Cutting & Deterding, 2024).

Application to Social Robots

Since the interactivity-as-demand model is fairly new, there is no extant research suggesting ways in which the model might apply to social robots. However, we can broadly speculate that the model calls for a bit of caution when considering where and how we might design social robots for human interaction. For example, we might suggest that *social robots need not engage all demand dimensions* to be effective towards myriad end goals. We see this already in the use of robots for therapy, using quasi-anthropomorphic robots designed to elicit emotional reactions (e.g., as social companions; see Cifuentes et al., 2020) might be built with a focus on emotional and social demands, but designed to be quite simple and easy to interact with (lowering cognitive and physical demand). This might be quite different than developing a social robot intended to present as increasingly human—and so, increasingly demanding. It could be that uncanny valley effects may be associated with unexpected increases in demand, such as increased negative affect (re: emotional demand) paired with increased cognitive dissonance (re: cognitive demands) as users try to sort out and categorize their robot companion (see Banks, 2021). The interactivity-as-demand model provides developers with four simultaneous concerns such that they could consider in comparison to the end-goal of the robot design and, from there, decide on a calibration of each source of demand relevant to achieving that goal.

CONCLUSION

Since at least the 1960s, video games and digital worlds have hosted within them variably social interactions between humans and NPCs—from basic artificial intelligences to other human users. Similarly, the development and introduction of social robots into various realities and corporealities closely mirrors these same dimensions. Social robots are programmed with various intelligences, up to and including control via human teleoperators. From a basic understanding of social presence and electronic propinquity, player-avatar relations and interactions, and the interactivity-as-demand model, we have visited established theory and data from wholly digital interactions that should allow for a more nuanced understanding of social robots. Although seamless human-robot interactions seem both novel and futuristic, we posit that decades of scholarship focused on digitality and interactivity in video games and digital worlds has given us a prescient and useful head-start for future development of social robots.

The cake just might be a truth, after all.[4]

REFERENCES

Banks, J. (2015). Object, me, symbiote, other: A social typology of player-avatar relationships. *First Monday, 20*(2).

Banks, J. (2021). Of like mind: The (mostly) similar mentalizing of robots and humans. *Technology, Mind, and Behavior, 1*(2).

Banks, J., & Bowman, N. D. (2015). From toy and tool to partner and person: Phenomenal convergence/divergence among game avatar metaphors. *AoIR Selected Papers of Internet Research, 5*.

Banks, J., & Bowman, N. D. (2016a). Emotion, anthropomorphism, realism, control: Validation of a merged metric for player–avatar interaction (PAX). *Computers in Human Behavior, 54*, 215–223.

Banks, J., & Bowman, N. D. (2016b). Avatars are (sometimes) people too: Linguistic indicators of parasocial and social ties in player–avatar relationships. *New Media & Society, 18*(7), 1257–1276.

Banks, J., Bowman, N. D., Lin, J.-H. T., Pietschmann, D., & Wasserman, J. A. (2019). The common player-avatar interaction scale (cPAX): Expansion and cross-language validation. *International Journal of Human-Computer Studies, 129*, 64–73.

Banks, J., & Carr, C. T. (2019). Toward a relational matrix model of avatar-mediated interactions. *Psychology of Popular Media Culture, 8*(3), 287–295.

---

4 In the 2007 video game *Portal*, players encounter an artificial intelligence embodied in a robot named GLaDOS (the game's only NPC), who promises rewards for solving in-game puzzles (such as cake). Early in the game, player come across graffiti on a hallway, "The cake is a lie." For those wanting to learn more, we won't spoil the game's message. =)

Banks, J., Koban, K., & Chauveau, P. de V. (2021). Forms and frames: Mind, morality, and trust in robots across prototypical interactions. *Human-Machine Communication, 2*, 81–103.

Bowman, N. D. (2016). Video gaming as co-production. In R. Lind (Ed.), *Produsing 2.0: The intersection of audiences and production in a digital world* (pp. 107–123). Peter Lang.

Bowman, N. D. (2021). Interactivity as demand: Implications for interactive media entertainment. In C. Klimmt & P. Vorderer (Eds.), *Oxford handbook of media entertainment*. Oxford University Press.

Bowman, N. D., & Banks, J. (2024). The [Object, Me, Symbiote, Other] in the machine: Insights from video game psychology for teleoperator-robot relations. *Proceedings of the Hawaiian International Conference on System Sciences, 57*, 610–619.

Bowman, N. D., Banks, J., & Downs, E. (2021). Mechanisms of identification and social differentiation in player-avatar relations. *Journal of Gaming & Virtual Worlds, 13*(1), 55–73.

Bowman, N. D., Keene, J. R., & Najera, C. J. (2021). Flow encourages task focus, but frustration drives task switching: How reward and effort combine to influence player engagement in a simple video game. *Proceedings of the CHI Conference Human Factors in Computing Systems* (no. 119). ACM.

Brandstätter, P., Sagmann, S., Krükel, D., Maurer, M., & Lankes, M. (2021). I will stand by you: Measuring the perceived social presence towards semi-autonomous companions in a 2D game. *Proceedings of the Annual Symposium on Computer-Human Interaction in Play* (pp. 5–9). ACM.

Cardona-Rivera, R., & Young, R. (2014). Games as conversation. *Proceedings of the AAAI Conference on Artificial Intelligence and Interactive Digital Entertainment, 10*(4), 2–8.

Chen, N., Liu, X., Zhai, Y., & Hu, X. (2023). Development and validation of a robot social presence measurement dimension scale. *Scientific Reports, 13*(1), 2911.

Chen, Y. C., Yeh, S. L., Lin, W., Yueh, H. P., & Fu, L. C. (2023). The effects of social presence and familiarity on children–robot interactions. *Sensors, 23*(9), 4231.

Cifuentes, C. A., Pinto, M. J., Céspedes, N., & Múnera, M. (2020). Social robots in therapy and care. *Current Robotics Reports, 1*, 59–74.

Cook, C. (2019). Between a troll and a hard place: The demand framework's answer to one of gaming's biggest problems. *Media and Communication, 7*(4), 176–185.

Cutting, J., & Deterding, S. (2024). The task-attention theory of game learning: A theory and research agenda. *Human–Computer Interaction, 39*, 257-287(

Edwards, A., Edwards, C., Westerman, D., & Spence, P. R. (2019). Initial expectations, interactions, and beyond with social robots. *Computers in Human Behavior, 90*, 308–314.

Ekman, I., Chanel, G., Järvelä, S., Kivikangas, J. M., Salminen, M., & Ravaja, N. (2012). Social interaction in games: Measuring physiological linkage and social presence. *Simulation & Gaming, 43*(3), 321–338.

Frommel, J., Dechant, M. J., & Mandryk, R. L. (2021). The potential of video game streaming as exposure therapy for social anxiety. *Proceedings of the ACM on Human-Computer Interaction, 5*, no. 258.

Gunkel, D. J., & Wales, J. J. (2021). Debate: What is personhood in the age of AI? *AI and Society, 36*, 473–486.

Haggadone, B. A., Banks, J,. & Koban, K. (2021). Of robots and robotkind: Extending intergroup contact theory to social machines. *Communication Research Reports, 38*(3), 161–171.

Häkkilä, J., Paananen, S., & Väänänen, K. (2022). Personalizing robot avatars—Opening the discussion. In K. Marky, U. Grünefeld, & T. Kosch (Eds.), *Mensch und Computer 2022— Workshopband MuC 2022*. Gesellschaft für Informatik e.V.

Hamdy, S., & King, D. (2017). Affect and believability in game characters: A review of the use of affective computing in games. In *Proceedings of the Annual Conference on Simulation and AI in Computer Games* (pp. 90–97). EUROSIS.

Heath, F. G. (1972). Origins of the binary code. *Scientific American, 227*(2), 76–83.

Hegel, F., Muhl, C., Wrede, B., Hielscher-Fastabend, M., & Sagerer, G. (2009). Understanding social robots. In *Proceedings of the International Conferences on Advances in Computer-Human Interactions* (pp. 169–174). IEEE.

Ho, J. C. F., & Ng, R. (2022). Perspective-taking of non-player characters in prosocial virtual reality games: Effects on closeness, empathy, and game immersion. *Behaviour & Information Technology, 41*(6), 1185–1198.

Holl, E., & Melzer, A. (2022). Moral minds in gaming: A quantitative case study of moral decisions in Detroit: Become human. *Journal of Media Psychology, 3*(5).

Hoorn, J. F. (2020). Theory of robot communication: I. The medium is the communication partner. *International Journal of Humanoid Robotics, 17*(6), 2050026.

Jo, D., Kim, K., Welch, G. F., Jeon, W., Kim, Y., Kim, K., & Kim, G. J. (2017). The impact of avatar-owner visual similarity on body ownership in immersive virtual reality. *Proceedings of the Symposium on Virtual Reality Software and Technology* (no. 77). ACM.

Junuzovic, S., Inkpen, K., Tang, J., Sedlins, M., & Fisher, K. (2012). To see or not to see: A study comparing four-way avatar, video, and audio conferencing for work. *Proceedings of the International Conference on Supporting Group Work* (pp. 31–34). ACM

Kang, H., & Kim, K. J. (2020). Private information disclosure on the Internet of Things: The effects of tailoring, self-expansion, and power usage. *Journal of Broadcasting & Electronic Media, 64*(4), 640–660.

Konami. (1997). *Castlevania: Symphony of the night*. [Video game]. Konami.

Korzenny, F. (1978). A theory of electronic propinquity: Mediated communication in organizations. *Communication Research, 5*(1), 3–24.

Kryston, K., Goble, H., & Eden, A. (2021). Incorporating virtual reality training in an introductory public speaking course. *Journal of Communication Pedagogy, 4*, 133–151.

Lee, K. M. (2004). Presence, explicated. *Communication Theory, 14*(1), 27–50.

Lin, J.-H. T., Wu, D.-Y., & Bowman, N. D. (2022). Beat Saber as virtual reality exercising in 360 degrees: A moderated mediation model of playable angles on physiological and psychological outcomes. *Media Psychology, 26*(4), 414–435.

Liszio, S., Emmerich, K., & Masuch, M. (2017). The influence of social entities in virtual reality games on player experience and immersion. *Proceedings of the International Conference on the Foundations of Digital Games* (no. 35). ACM.

Lombard, M., & Ditton, T. (1997). At the heart of it all: The concept of presence. *Journal of Computer-Mediated Communication, 3*(2), JCMC321.

Minsky, M. (1980, June). Telepresence. *OMNI Magazine*, 44–52.

Montalvo, F. L., Alves, G. M., Payne, C. A., Sasser, J. A., McConnell, D. S., & Smither, J. A. (2022). Trait loneliness and social presence in human-human and human-robot interaction. *Proceedings of the Human Factors and Ergonomics Society Annual Meeting, 66*(1), 817–821.

Nintendo. (1986). *The Legend of Zelda*. [Video game]. Nintendo.

Pereira, A., Prada, R., & Paiva, A. (2014). Improving social presence in human-agent interaction. In *Proceedings of the SIGCHI Conference on Human Factors in Computing Systems* (pp. 1449–1458). ACM.

Pollen Robotics. (n.d.). *Reachy*. Retrieved from <https://www.pollen-robotics.com/reachy/>

Pressgrove, G., & Bowman, N. D. (2021). From immersion to intention? Exploring advances in prosocial storytelling. *International Journal of Nonprofit and Voluntary Sector Marketing, 26*(2), e1689.

Rafaeli, S., & Sudweeks, F. (1997). Networked interactivity. *Journal of Computer-Mediated Communication, 2*(4), JCMC243.

Sajjadi, P., Hoffmann, L., Cimiano, P., & Kopp, S. (2019). A personality-based emotional model for embodied conversational agents: Effects on perceived social presence and game experience of users. *Entertainment Computing, 32*, 100313.

Schumann, C., Bowman, N. D., & Schultheiss, D. (2016). Quality in video games: Subjective quality assessments as predictors of self-reported presence in first-person shooters and role-playing games. *Journal of Broadcasting & Electronic Media, 60*(4), 547–566.

Short, J., Williams, E., & Christie, B. (1976). *The social psychology of telecommunications*. Wiley.

Steuer, J. (1992). Defining virtual reality: Dimensions determining telepresence. *Journal of Communication, 42*(4), 73–93.

Tamborini, R., & Bowman, N. D. (2010). Chapter 5: Presence in video games. In *Immersed in media* (pp. 105–128). Routledge.

Topal, S., Erkmen, İ., & Erkmen, A. M. (2010). Towards the robotic avatar: An extensive survey of the cooperation between and within networked mobile sensors. *Future Internet, 2*(3), 363–387.

TVTropes.org. (n.d.). *Story difficulty setting*. TVtropes.org. Retrieved from <https://tvtropes.org/pmwiki/pmwiki.php/Main/StoryDifficultySetting>

Valve. (2007). *Portal*. [Video game]. Valve Corporation

Walther, J. B., & Bazarova, N. N. (2008). Validation and application of electronic propinquity theory to computer-mediated communication in groups. *Communication Research, 35*(5), 622–645.

Wellenreiter, M. (2015). Screenwriting and authorial control in narrative video games. *Journal of Screenwriting, 6*(3), 343–361.

Yoon, B., Kim, H., Lee, G. A., Billinghurst, M., & Woo, W. (2019). The effect of avatar appearance on social presence in an augmented reality remote collaboration. *Proceedings of the Conference on Virtual Reality and 3D User Interfaces* (pp. 547–556). IEEE.

CHAPTER SEVENTEEN

Cuteness & Repulsiveness: Aesthetics of Machine Bodies

JOEL GN

Situated at the intersection of form and function, social robotics presents questions concerning the anthropomorphism and acceptance of artificial agents. To a significant extent, the experience of humanizing an object that simulates but fundamentally differs from a human companion marks a departure from earlier imperatives, where robots were for the most part built to automate manufacturing processes. As machines that solicit human affection, social robots (as the adjective "social" denotes) are approachable *and* interactive agents that are indeed capable of more than just mechanical gestures (Gn, 2017).

In other words, the apparent success of a social robot does not solely depend on its internal machinery, but also on the relationship fostered between the human subject and the robotic object. When the Sony AIBO robotic dog was temporarily discontinued in 2006, owners and Sony engineers continued to provide repair services until the supply of component parts was exhausted (Mochizuki & Pfanner, 2015). Several owners even conducted funeral rites, when no other avenue to "revive" the pet was available (Brown, 2015). These incidents point to the pertinence of a specific interactive context that allows the subject to bypass a more realistic understanding (i.e., that it is, in fact, a machine) for an anthropomorphic connection with the object. As Airenti notes, anthropomorphism places non-human objects as "interlocutors in a communicative interaction," resulting in the "automatic attribution of intentionality and social behaviour" (Airenti, 2018, p. 10). This notion is broadly aligned with Duffy's weak AI approach, which posits the social robot should sustain "a balance of illusion that leads the user to believe in the sophistication of the system in areas where the user will not encounter its failings" (Duffy, 2003, p. 178). From a design standpoint, this balance of illusion underscores a tension between anthropomorphism and brute machinery; for even as the latter is indispensable for the social robot's operation, it also ought to be sufficiently concealed for the user to regard the robot as something more than a machine.

At the same time, however, anthropomorphism does not neatly translate into acceptance. According to Duffy's rationale, human likeness can be detrimental to the human-robot relationship (e.g., the Uncanny Valley), given how users would inevitably compare the social robot to an actual human being (2003). To look but *not* behave like a human would in this sense, exacerbate rather than downplay the robot's machine-based characteristics. Responding to this issue of acceptance, the experience of cuteness becomes a potent design solution, insofar as it elicits responses of care and affection within the interactive space. This acceptance, I contend, raises a stark contradiction: In a milieu where new technologies are extensively simulating human attitudes and behaviors, animated characters, toys, and digital avatars are accepted *in spite of* their visible artificiality. The application and appreciation of cute design in social robotics thus brings this contradiction to the fore by (1) using abstracted human/non-human forms, and (2) shifting the function of machines as instruments of labor to objects of emotional investment.

DEPENDENCE, SOCIALITY AND SIMPLICITY

Given the contribution of cuteness to social robotics, what are the essential features of its design and how does it bring about an intimate connection between a human subject and robotic object? According to the Oxford English Dictionary, "cute" is an adjective primarily used to describe a thing that is attractive and endearing. While this definition is empirically dependent on sociocultural nuance, it is still possible to observe three interconnected domains within the aesthetic. First, the cute body is often (but not always) recognized for its rounded, infantilized features and even clumsy demeanor that places it in a position of dependence. Robots aside, these characteristics are often found in human infants and the young of other mammals like dogs, cars, or bears. On the surface, our attraction to such objects seems to support the hypothesis that "the recognition of the specialness of the young" provides a species with the advantages for survival (Morreall, 1991, pp. 39–40). No doubt an evolutionary approach would conceive of cuteness as an ethological factor necessary for parent-child attachment, but it overlooks the pertinent incongruence in appearance between infants and adults. While infants are not a different species, they nevertheless possess a smaller, chubbier body structure and are generally perceived as clumsy inferiors. On this basis, one could reasonably claim that our attachment to infants—and by association, cute objects—is attributed to a non-repulsive deformity of the more independent and functional adult body.

Second, cuteness stimulates sociality insofar as the subject can easily obtain gratification through a connection with the cute object. In the context of social robotics, this gratification pertains to both the affection and attachment experienced by the human for the robot; it points to a level of sociality that is fundamentally different from the parent-child attachment, especially when one considers how children themselves anthropomorphize toys and other images. This proposition is supported by the work of Sherman and Haidt (2011) who observe that cuteness is a mechanism of sociality that is equally relevant in situations that do not involve parental affection and caregiving. Following this approach, the anthropomorphism within cute design can be understood as a calibrated experience that communicates an intentionality understood in relatively "human" terms. Referring to such phenomena as an outcome of the *intentional stance*, Dennett (1987) writes:

> The intentional stance provides a vantage point for discerning similarly useful patterns. These patterns are objective—they are there to be detected—but from our point of view they are not out there entirely independent of us, since they are patterns composed partly of our own "subjective" reactions to what is out there; they are patterns made to order for our narcissistic concerns. (p. 39)

Similar to how humans anthropomorphize or make sense of a non-human object, the intentional stance is often adopted when information on the object's exact properties and functions are not available or difficult to comprehend. The obvious lack of an objective perspective, in this instance, does not undermine the utility of cuteness. On the contrary, I contend that cute design provides a repertoire of concrete strategies for engaging with and making sense of the artificiality of social robots.

Third, cuteness takes advantage of simplicity to design and develop objects that are potentially open to interlocutors' interpretation. In a case study of the iconic Hello Kitty, Brian McVeigh (2000) notes that the character's plainness "characterises her as a cryptic symbol waiting to be interpreted and filled with meanings" (p. 234). Compared to the complexity of painting, architecture, or even circuit boards, the stark lack of detail in cute design becomes an avenue of appropriation for the generic user. This approach to form also corresponds to more contemporary cultural perceptions of machine bodies, which have ostensibly shifted from mechanical, differentiated complexity to a more organic and unified form. As scholars such as Black (2014) and Kakoudaki (2014) observe, robots and automata before the 20th century were depicted with multiple parts and were interpreted as images of order, predictability, and de-individuation. In contrast, the modern gadget is conventionally organized around principles of unity and coherence. Social robots, tablets, mobile phones, and wearables, for example, possess a minimalist exterior concealing any machine-based complexity and are primarily assessed for their overall user experience rather than the functions of specific mechanisms. If an anatomical mode of vision once treated the body as a *Gestalt* of clearly defined parts, then current conceptions as exemplified by cute design arguably frame the body as a conduit of data and information. This pervasive expression of the aesthetic within product design suggests cuteness is not simply construed as a felt experience or a tendency to anthropomorphize, rather it is employed as a stylistic device with its own vocabulary for fostering an affectionate subject-object relationship, within a given sociocultural space.

Taken together, the domains of dependence, sociality, and simplicity constitute a paradigm for cuteness that can be implemented in a variety of contexts pertaining to the design and acceptance of social robots. It is, therefore, not uncommon for designers to use culturally specific sensibilities of cuteness to build and design objects for the target market, since cute design can act as a template for appearances, attitudes, and behavior that would otherwise be difficult to describe and express.

ATTRACTION AND REPULSION

Reducing the perceived distance between human and robot with cute design can occur at two levels. The first builds upon the intentional stance and anthropomorphizes the robot by ascribing intentions within the interaction. On the one hand, this anthropomorphism entails a tangible suspension of disbelief whereby the human negates the non-humanness of the object in order to participate in the interaction. Disturbances to such an interaction will then, in part, depend on how the object participates and performs its role within the interaction. On the other hand, the anthropomorphic relation, as argued by Airenti (2018), is not necessarily cooperative. An object can be anthropomorphized as an obstacle, competitor, or even antagonist, which arguably reduces acceptance and increases the social distance between human and robot. Evaluating the success of a social robot based on the presence of an anthropomorphic relation would not be helpful since the sentiments of the human user are not considered.

Second, cute design augments a positive anthropomorphism, which goes beyond ascribing human intentions to appearances and gestures that are effective for human-robot attachments. I would, however, stress that this augmentation is not without ambiguity. As amply illustrated by prominent products in the field, mixing cuteness with robotics does not result in a photorealistic representation of the human or animal form, even though these objects simulate pro-social and "human" behaviors. By working with and leveraging cues that motivate positive attachment, cute design is an abstraction that selectively reduces the social distance between the human user and the social robot.

But how are these cues identified, and do they actually make the robot seem more "human?" One viable response can be obtained from Haslam's (2006) in-depth inquiry on dehumanization. Making a clear distinction between characteristics that are uniquely human (UH) and those that are intrinsic to human nature (HN), he explains that objects lacking UH characteristics tend to be regarded as animal-like, while those without HN characteristics would be deemed as mechanistic. Both these categories are connected to two types of dehumanization, namely animalistic (emotions and gestures that resemble animals) and mechanistic (persons and things that behave like machines). Haslam's exercise is comprehensive, and it is worth noting that certain traits labeled as dehumanizing in the study (e.g., child-likeness, passivity) are also found in some cute objects. By avoiding biological realism and conventional adult traits, cute design fabricates a stylized caricature of human attitudes and behaviors that, sans the pleasure of the human interlocutor, makes the social robot more deformed yet less repulsive. As eloquently described by Harris (1992):

> The aesthetic of cute creates a class of outcasts and mutations, a ready-made race of lovable inferiors... Something becomes cute not necessarily because of a quality it has but a quality it lacks, a certain neediness and inability to stand alone. (p. 179)

Although the claim draws out the contradiction of the cute object's mutation and dependence, Harris also attends to the ambiguity that social psychology approaches do not consider. That is, the acceptability of cuteness continues to present a deformity that facilitates the human's control over the non-human object. Taking the point further, social robots like NEC's PaPeRo bear no likeness to the human body; instead, they are a cute abstraction that "deforms" the human body with their smooth, rounded, and minimalist exterior. In augmenting dependence, sociality, and simplicity, the paradigm of cuteness presents highly stylized, sanitized bodies that are effectively uncoupled from repulsive human characteristics. At first glance, there seems to be a convenient convergence between cute design and Dennett's (1987) view that the intentional stance caters to our narcissistic concerns, but I propose that cute design—particularly in the field of social robotics—reflects deeper anxieties that question and challenge the boundaries of the familiar and strange.

AN UNCANNY LIKENESS OF BEING

So why is the omission of certain human characteristics beneficial for a social robot's acceptance? For Mori (1970), an object will stimulate greater acceptance if it appears and acts like a human; however, there is a region in the design space, known as the Uncanny Valley, where such attributes will come across as strange and even repulsive. While Mori's model has often been cited to describe the peculiar connection between a robot's appearance and acceptance, it falls short of demonstrating how any sort of realist verisimilitude can be uncanny in the first place. One can certainly excuse robotic researchers for not speculating further given the model's chief concern of positive human-robot interaction, but a deeper investigation into the implications of the uncanny is required if we are to touch on the sense of likeness (or lack thereof) assumed by social robots.

One helpful starting point can be found in Freud's (2003) influential work, predictably titled *The Uncanny*, where the word is defined as "what evokes dread and horror… so it commonly merges with what excites fear in general" (p. 123). Yet at the same time, the uncanny is also concerned with things or experiences that are familiar, which raises the following question: How can something familiar to us also be frightening, dreadful, and hence repulsive? By considering the uncanny along with the familiar, Freud delves into the polysemy of the German *heimlich* to refer to a familiar situation or object that is "familiar, tame, intimate, friendly" (p. 126). Like the word "cute," *heimlich* is often used in a positive context, such as the human's attachment to an object that offers comfort and security. Conversely, the other meaning of *heimlich* is more enigmatic, as it refers to a thing that is "concealed, kept hidden, so that others do not get to know…" (p. 129). This other idea of *heimlich* as secrecy draws on its association with the antonym *unheimlich*, which describes that which is eerie, weird, or arouses fear. In Freud's writing, the familiar is not necessarily an object or place that we can eagerly approach, because it may be filled with hidden, spectral qualities that haunt and strike fear.

There are, in my view, a couple of insights that can be gathered from this etymological theorization of the uncanny. First, the uncanny, in its likeness, stimulates fear and anxiety, whereas cute design circumvents these sentiments by humanizing the artificial, and not replicating the biological. The resulting deformation, in this sense, is not repulsive, but reinforces the affection expressed by the human user. To use another example, Toyota's Kirobo Mini is a humanoid robot that would unequivocally be considered cute—lacking any visible feature that causes disgust or discomfort, it is neither endowed with the facial features nor the anatomy of an adult human being. As an inorganic, simplified animation, the body and behavior of the Kirobo is presented as a lovable humanization and not a realistic human.

Second, cute design has so far been understood as a solution to the problem of the uncanny via an abstraction of a lovable other, but this abstraction is once again placed within a contradiction, for what is considered "acceptable" in the social robot is far from the actuality of the human but instead is the systematically humanized difference of the machine. This does not mean that cuteness is immutably antithetical to the uncanny; on the contrary, it works from and depends on one's understanding of the uncanny so as to be distinct from it (Gn, 2017). Considering how product designers are made aware of the characteristics and flaws that make a social robot unacceptable to a human, the modifications engendered by cute design do not nullify the effects of the uncanny but are indirectly cognizant of them. To be deformed by design, in other words, is to bring about an acceptance based on this acquired difference of what is approachable and repulsive.

INDUSTRIAL ATTACHMENTS

In a market where physical and emotional experiences are increasingly being commodified, social robotics has made remarkable contributions to the surge of emotional media that rapidly closes the gap between technology and human affection. But even as they have been touted as replacements for human presence, current designs and user experiences are evidently part of a more rigid process of industrialization that tends to *repeat*, rather than innovate from established impressions of cuteness. If the uncanny provokes revulsion, the cuteness of social robots thoroughly defuses it through systematization. In the interaction with the social robot, the human is not presented with the complex, delicate (and at times, unsettling) difference of personhood, but with a machine operating from a static abstraction of relationality.

Yet as I have sought to demonstrate, cute design continues to be a means for us to come to terms with the object's difference, because it enables us to get close to and develop an attachment to the object. This has provoked questions on the nature and mechanisms of our social predispositions, and how popular aesthetics such as cuteness serve to reflect such desires and anxieties. Perhaps, the uncanny is not a neatly defined space for what we should avoid, but a permanent site of contention for us to reframe and reconsider prevailing notions of the cute and lovable? Any study on the appeal of social robots should, in this regard, recognize that our attraction to them is concurrently compelled by our impulse to connect with the other and hindered by what is disconcertingly familiar about ourselves.

On this note, it is difficult to determine if the artificiality of social robots inevitably makes human-robot relationships less authentic than what is shared between humans; after all, the positive anthropomorphism of social robots has already presented alternative visions of who or what can be accepted and even loved. The more pertinent concern, I believe, is to actively consider an approach that persists in meaningful engagement with the messy and apparently undesirable facets of the uncanny. For, if cuteness is indeed the context and contradiction necessary to resolve the difference of the machine, then a critique on the imagined certainties (and designs) of robotic embodiment would continue to be worthwhile. The social robot may for a time sustain the illusion that our interpersonal uncertainties can be automatically resolved and reined in, but its only in the valley of the uncanny that we learn to confront and possibly accept what we have been avoiding all along.

REFERENCES

Airenti, G. (2018). The development of anthropomorphism in interaction: Intersubjectivity, imagination and theory of mind. *Frontiers in Psychology, 9*, 1–13.

Black, D. (2014). *Embodiment and mechanization: Reciprocal understandings of body and machine from the renaissance to the present*. Ashgate.

Brown, A. (2015, March 12). To mourn a robotic dog is to be truly human. The Guardian. <https://www.theguardian.com/commentisfree/2015/mar/12/mourn-robotic-dog-human-sony>

Dennett, D. C. (1987). *The intentional stance*. MIT Press.

Duffy, B. R. (2003). Anthropomorphism and the social robot. *Robotics and Autonomous Systems, 42*(3–4), 177–190.

Freud, S. (2003). *The uncanny* (D. McLintock, Trans.). Penguin.

Gn, J. (2017). Designing affection: On the curious case of machine cuteness. In J. P. Dale, J. Leyda, J. Goggin, A. P. McIntyre, & D. Negra (Eds.), *The aesthetics and affects of cuteness*. Routledge.

Harris, D. (1992). Cuteness. *Salmagundi, 96*, 177–186.

Haslam, N. (2006). Dehumanization: An integrative review. *Personality and Social Psychology Review, 10*(3), 252–264.

Kakoudaki, D. (2014). *Anatomy of a robot: Literature, cinema, and the work of artificial people*. Rutgers University Press.

McVeigh, B. J. (2000). How Hello Kitty commodifies the cute, cool and camp: "Consumutopia" versus "control" in Japan. *Journal of Material Culture, 5*(2), 225–245.

Mochizuki, T., & Pfanner, E. (2015, February 11). In Japan, dog owners feel abandoned as Sony stops supporting "Aibo." *The Wall Street Journal*. <https://www.wsj.com/articles/in-japan-dog-owners-feel-abandoned-as-sony-stops-supporting-aibo-1423609536>

Mori, M. (1970). The uncanny valley. *Energy, 7*(4), 33–35.

Morreall, J. (1991). Cuteness. *British Journal of Aesthetics, 31*(1), 39–47.

Sherman, G. D., & Haidt, J. (2011). Cuteness and disgust: The humanizing and dehumanizing effects of emotion. *Emotion Review, 3*(3), 245–251.

CHAPTER EIGHTEEN

Sex & Gender: A Complicated Relationship

LEOPOLDINA FORTUNATI

The history of technology shows that the forms of machines often emulate certain human body components: Even the airplane seems to have a kind of head (the nose), arms (the wings), and a trunk (the fuselage), since our brains are in a sense "programmed" to see the human in the nonhuman, as some suggest (e.g., Dacey, 2017). But no other technology has been submitted to the deep process of anthropomorphism experienced by robots during their long evolution. The 20th century has propelled this anthropomorphism process significantly, resulting for example in the creation of near-flawless androids (e.g., Nishio et al., 2007). Concurrently, awareness of the potential hazards linked to robot anthropomorphization has grown, thanks to considerations of the uncanny valley (Mori, 1970). This concept has reignited debates surrounding the authenticity of social robots that simulate human-like characteristics but remain inherently distinct. While the rationale behind this protracted process of robot anthropomorphization has yet to be fully clarified by roboticists, a substantial portion of their efforts has continued to be invested in replicating human bodily appearances.

Elsewhere (Fortunati, 2013), I have proposed a rationale for the anthropomorphism of robots. I hypothesized that this persistent process might stem from men's envy of women's unique capacity to give birth, an attempt to diminish women's biological monopoly by creating artificial entities resembling humans, such as robots. Regardless of the interpretation of this process, its complexities warrant careful investigation. The current research landscape presents conflicting findings concerning user preferences for various forms of social robots. Some studies show a preference by users for humanoid robots (e.g., Guo et al., 2022), while others demonstrate a preference by them for machine-like robots (e.g., Vlachos et al., 2016).

Selecting an anthropomorphic form, as opposed to zoomorphic, phytomorphic, or mechanomorphic shapes, prompts an immediate decision: Determining the type of human form to emulate. Human diversity renders an undifferentiated human form non-existent, demanding choices not only regarding the type of humans to replicate but also the manner of replication—stylized or realistic. Unlike other mediums of visual human representation, such as painting and sculpture, social robots presenting anthropomorphic appearances risk perpetuating stereotypes related to gender, race, and aging. Artistic forms often transform and reveal inherent or latent truths in visual representations, whereas social robots, driven by utilitarian and functional purposes, lack this transformative capability.

Anthropomorphized social robots are susceptible to replicating prevailing cultural gender norms, like Sophia (Hanson Robotics) who presents as a thin, White female who giggles, reinforcing norms around body size, the Whiteness of technology, and diminutive behavior. These norms, predominantly binary in nature (male and female), do not encapsulate the multifaceted complexity of contemporary gender identities, which are complex and sometimes fluid, including people identifying as genderqueer, gender fluid, and transgender. Rakow (1986) argues that "gender is both something we do and something we think with, both a set of social practices and a system of cultural technologies. The social practices—the 'doing' of gender—and the cultural meanings—'thinking the world'— … constitute us" as gendered individuals (p. 21). Against this dynamic definition of gender, assuming a binary vision of gender more aligned with sex (i.e., the physical, observable attributes of bodies) may lead robot designers to inadvertently stereotype gender attributes. Scholars like Alesich and Rigby (2017) question this binary gender perspective, which dominates robotics discussions. Genderless categorizations, applied to machines, provide a counterpoint to this binary vision. Moreover, a study from Banks and Koban (2022) suggests that robots can have gender and race cues and still not be gendered or racialized in perception, confirming the finding obtained by Fortunati et al. (2022).

Alesich and Rigby (2017) prompt reflection on the implications of creating gendered robots for human understanding of gender. Scholarly perspectives converge on the idea that the more humanlike a robot becomes, the more it embodies sexual identity. Some researchers like Bray (2012) and Søraa (2017) acknowledge that language and communication make gendering inevitable because pronouns and naming conventions can further solidify gender associations. Crowell et al. (2009) extend this discourse, asserting that not only do anthropomorphizing language and sex-cued bodies, but also the mere embodiment of robots introduces questions about sexual identity. Specifically, they found that male-embodied and female-disembodied machine voices are perceived as more reliable, while embodied robots in general are perceived as friendlier.

Thus far, roboticists have concentrated primarily on giving social robots gendered forms, with minimal focus on representing different races or ages beyond the default white model. The discourse on aging introduces further complexity, as robots like Pepper and NAO display features reminiscent of young boys, though without a clear conceptualization. It is unsurprising that discussions about the appearance of robots have predominantly centered around gender, while limited literature addresses race and even less delves into age. Existing gender and social robot research has explored robot gender, gender attitudes, and behaviors exhibited by potential users towards robots, chatbots, and virtual assistants (Fortunati & Edwards, 2022; Fortunati, 2023). In the upcoming sections, both approaches will be briefly examined.

GENDERING CUES AND HUMAN PERCEPTION

Regrettably, a paucity of studies exists regarding the rationale and decision-making process of roboticists in gendering robots. An exception is Masterson (2022), who scrutinized how CEO Matt McMullen, the visionary and creator of RealDoll, crafted his perspective on the Harmony model. McMullen's motivations behind the RealDoll model were elucidated in interactions with the public and media. Through critical discourse analysis of the marketing and publicity surrounding this sex robot, Masterson unveiled the ambiguity within McMullen's discourse. It oscillates between categorizing Harmony as a mechanical artifact and a work of art versus a companion simulating or approximating human characteristics. Fortunati and Edwards (2022) emphasize how creators and marketers strategically exploit this technological ambiguity to serve the interests of developers and financial stakeholders, often to the detriment of larger discussions on the societal and political implications of these technologies (p. 33) and to the detriment of current gender diversification.

In general, most studies pertaining to the gender of machines contribute to theoretical debates on a socio-political level through research papers or discussions. Robertson (2010), in a discussion paper, posits that, unlike humans, robots are generally gendered even in the absence of visible physical genitals, which often trigger gender attribution processes in humans. However, Jung et al. (2016) documented that minimal visual gender cues on a robot's body suffice for people to ascribe gender to it. They also found that in the absence of gender cues, there's a prevalent tendency to perceive robots as male by default. When cues are introduced, a robot with a male cue is perceived as more masculine, while one with a female cue is perceived as more feminine.

According to Nass et al. (1997), the voice constitutes the primary cue triggering gender stereotyping of machines. Evaluations provided by a computationally male voice are received more seriously than those from a computationally female voice. This finding was subsequently corroborated by Powers and Kiesler (2006). These research outcomes underscore the voice's reflection of power stratification at the societal level. Nass and Yen (2010) recounted an incident wherein BMW had to recall one of its car models due to German male drivers rejecting directions from an acoustically female voice simply because it "was a woman." However, with women's growing societal influence, this narrative changed; within a few years, a majority of GPS navigators in the European Union featured female voices.

While the voice remains a central cue, other peripheral indicators such as hair length or lip color (e.g., pink versus gray) are often employed to gender robots. Given that robots are designed to act on behalf of or assist humans in various tasks, their gendered forms are often shaped by specific professions. This aspect also extends to the stereotypical gender associated with the programmed occupational role. Tay et al. (2014) discovered that participants in their research favored gender-occupational role stereotypes that aligned with personality-occupational role stereotypes, responding to robots in line with societal models. They also revealed that stereotyped personality might outweigh gender stereotypes in interactions with social robots, as the former mitigates the impact of the latter on user responses. Additionally, Bryant et al. (2020) noted that perceived occupational competence might be more influential than gender stereotypes, as it "is a better predictor for human trust than robot gender or participant gender" (p. 13).

The matching hypothesis, wherein a robot's appearance aligns with stereo typical occupational roles, suggests that this can increase user compliance with the robot (e.g., Carpenter et al., 2009). However, it also implies that such alignment could reinforce gender and occupation stereotypes, perpetuating gender disparities in society. For example, if a female-cued robot functions as a nurse, it is likely to perpetuate societal assumptions that nurses must be women, while in reality that social roles are gender-nonspecific. Powers et al. (2005) contend that challenging occupational stereotypes in robot design, such as creating a female mechanic's helper robot, might require more effortful communication for the human to make sense of the a-stereotypical robot.

Efforts have been made to challenge gender stereotypes. Eyssel and Hegel (2012) advocated for the development of gender-neutral or counter-stereotypical machines to counteract the persistence of personal and cultural stereotypes. Tam and Khosla (2016) proposed personalization as a strategy to dismantle gender stereotypes in robot appearances, however it is possible that people would simply personalize robots according to familiar tropes. Discourses on gendered stereotypes and social robots have benefited from two sources of inspiration that are beyond summarizing in this short chapter but that I recommend for further reading on the topic. The first is the Stanford University website (n.d.) encouraging a reconsideration of gender norms through various strategies and inviting designers to create robots that promote social equality. The second source is the European-funded research project GEECCO on gender and feminist aspects in robotics; Pillinger (2019) conducted an insightful analysis of key feminist contributions to robotics, addressing pivotal issues raised by feminist philosophers, theorists, and sociologists, thus advancing the discourse on gendered robots.

INTERPLAYS BETWEEN HUMAN GENDERS AND ROBOT GENDER CUES

Within this section, I delve into users' responses to gendered robots. Initially, the overall outcomes from numerous studies concerning the attitudes of both men and women toward robots appear to be conflicting. For instance, Taipale and Fortunati (2018), using a representative sample of Europeans ($N = 26,751$), demonstrated that among individuals expressing fairly or very positive views about robots, men slightly outnumbered women. Nomura (2017) also found that women participants in her research were less likely to hold a positive outlook on robots compared to their male counterparts (although she underscored the potential interaction of gender differences with moderating factors). However, other investigations (e.g., Reich-Stiebert & Eyssel, 2017) did not identify significant differences in attitudes expressed by men and women towards robots.

The contradictions in these findings can be attributed to the novelty of social robots that are primarily still in a prototyping phase (e.g., in university laboratories), leading to limited direct exposure and experience among individuals. Additionally, the research on gender and social robots has primarily originated from experiments conducted by engineers, computer scientists, and psychologists, often with small convenience samples (frequently students). Even the research questions are typically those that can be submitted to an experimentation process and thus are very much circumscribed. Moreover, the stimuli are in many cases pictures or videos of robots rather than live social robots, which mean findings are about *depictions of* robots rather than direct experiences of them. Given these research characteristics, the resulting findings are descriptive and not widely generalizable. Furthermore, studies into gender attitudes toward robots typically assume that participants will interpret gender cues corresponding to the intentions with which the researcher presented them. Many experiments involve a small number of participants who are asked about the gender they attribute to specific robots, and the options are often limited to male or female. However, when exploring robot gender through open-ended questions or free association exercises, we (Fortunati et al., , 2022) demonstrated that respondents constructed largely disembodied or ungendered (or only implicitly gendered) images of the social robot (Sophia) or the semi-robot (Alexa), despite the presence of powerful gender cues like a feminine persona (voice, personality, behaviors) (also observed by Humphry & Chesher, 2021).

Research on human sex differences in human-robot interaction draws from the CASA (Computers as Social Actors) paradigm (e.g., Reeves & Nass, 1996), and Eyssel and Hegel (2012) expanded this framework to robotics. They emphasized that robots' appearance, voice, and behavior provide social cues guiding social perception and categorization processes. A more limited exploration has focused on the intersection of human and robot gender, and even in these cases, results remain inconsistent. Otterbacher and Talias (2017) demonstrated that men have more affective reactions to female-cued robots than women are to male-cued robots. Eyssel et al. (2012) on the contrary identified a preference for robots of the same gender based on voice-gender cues. Siegel et al. (2009) found that participants rated robots of the opposite sex as more credible, trustworthy, and engaging, particularly among male subjects and female robots. Schermerhorn et al. (2008) noted differences in how men and women conceptualize, react to, and interact with robots; men tend to perceive robots as more human-like, to perform tasks better in robots' presence, and express more socially to a survey administered by robots, all compared to women interacting with robots. Kuchenbrandt et al. (2014) revealed that when people were instructed by either a "male" or a "female" robot to perform tasks stereotypically associated with females (sorting sewing equipment into a sewing box), they made more errors compared to tasks stereotypically associated with males (sorting different tools into a toolbox). Additionally, participants interacting with robots within stereotypically female work tasks were less inclined to accept help from robots than those interacting in stereotypically male tasks. A surprising finding was that within stereotypically female tasks, male and female robots were equally perceived as competent, whereas within stereotypically male tasks the female robot was deemed less competent than the male counterpart. These results partly contradict previous findings suggesting that people prefer a gender match between robots and task features.

TOWARD MORE WOMEN ....

In this chapter, I aimed to explore the multifaceted discourse surrounding gender and social robots. Given the emerging nature of social robots, this debate is still in its infancy, which accounts for the contradictory findings that have emerged in these initial stages of research. From our discussion it emerges clearly that the gendering of robots is a minefield. It is so difficult and complicated to unravel a non-stereotypical design of robots that maybe it should be better to focus on mechanomorphic or zoomorphic or phytomorphic robots—looking like typical machines, animals, or plants. What is certain, however, is that there are still too few women in charge of this sector. Robotics suffers from an insufficient presence of powerful women among designers, roboticists, and computer scientists. Without a critical mass of women capable of reconnecting the prevalent philosophical perspectives, theoretical approaches, and research methodologies to the history, visions, points of view, and interests of women, the world of robotics will remain heavily masculine. It is not enough to build non-stereotypically gendered robots to be truly inclusive of women. It is necessary to include women in the ideation process, then in the design process, and in the implementation process, and so on. Only by doing so we will be able to build robots that will respond really to the needs and desires of women. While the approach of participatory design has understood this need very well, the problem is that in the overall sector the practice remains very far from theory. So, the artist Martina Mara is right in reiterating this concept in her manifesto: "I'm for more women in robotics, not for more female robots!" (Hieslmair, 2017).

REFERENCES

Alesich, S., & Rigby, M. (2017). Gendered robots: Implications for our humanoid future. *IEEE Technology and Society Magazine, 36*(2), 50–59.

Banks, J., & Koban, K. (2022). A kind apart: The limited application of human race and sex stereotypes to a humanoid social robot. *International Journal of Social Robotics, 15*, 1949–1961.

Bray, F. (2012). Gender and technology. In *Women, science, and technology: A reader in feminist science studies* (pp. 370–381). Routledge.

Bryant, D., Borenstein, J., & Howard, A. (2020). Why should we gender? The effect of robot gendering and occupational stereotypes on human trust and perceived competency. In *Proceedings of the International Conference on Human-Robot Interaction* (pp. 13–21). ACM.

Carpenter, J., Davis, J. M., Erwin-Stewart, N., Lee, T. R., Bransford, J. D., & Vye, N. (2009). Gender representation and humanoid robots designed for domestic use. *International Journal of Social Robotics, 1*, 261–265.

Crowell, C. R., Villano, M., Scheutz, M., & Schermerhorn, P. (2009). Gendered voice and robot entities: Perceptions and reactions of male and female subjects. In *Proceedings of the International Conference on Intelligent Robots and Systems* (pp. 3735–3741). IEEE.

Dacey, M. (2017). Anthropomorphism as cognitive bias. *Philosophy of Science, 84*(5), 1152–1164.

Esposito, A., Amorese, T., Cuciniello, M., Riviello, M. T., Esposito, A. M., Troncone, A., ... & Cordasco, G. (2019a). Elder user's attitude toward assistive virtual agents: The role of voice and gender. *Journal of Ambient Intelligence and Humanized Computing, 12*, 4429–4436.

Eyssel, F., & Hegel, F. (2012). (S)he's got the look: Gender stereotyping of robots. *Journal of Applied Social Psychology, 42*(9), 2213–2230.

Eyssel, F., de Ruiter, L., Kuchenbrandt, D., Bobinger, S., Hegel, F. (2012). "If you sound like me, you must be more human": On the interplay of robot and user features on human robot acceptance and anthropomorphism. In *Proceedings of the International Conference on Human-Robot Interaction* (pp. 125–126). ACM.

Fortunati, L. (2013). Afterword: Robot conceptualizations between continuity and innovation. *Intervalla: Platform for Intellectual Exchange, 1*, 116–129.

Fortunati, L., & Edwards, A. P. (2022). Gender and human-machine communication: Where are we? *Human-Machine Communication, 5*, 7–47.

Fortunati, L. (2023). Gender and identity: Some notes. In R. McEwen, A. Guzman & S. Jones (Eds.), *The SAGE handbook of human-machine communication* (pp. 127–135). SAGE.

Fortunati, L., Edwards, A., Manganelli, A. M., Edwards, C., & de Luca, F. (2022). Do people perceive Alexa as gendered? A cross-cultural study of people's perceptions, expectations, and desires of Alexa. *Human-Machine Communication, 5*, 75–97.

Guo, F., Li, M., Chen, J., & Duffy, V.G. (2022). Evaluating users' preference for the appearance of humanoid robots via event-related potentials and spectral perturbations. *Behaviour & Information Technology, 41*(7), 1381–1397.

Hieslmair, M. (2017, March 8). Martina Mara: More women in robotics. *Ars Electronica*. <https://ars.electronica.art/aeblog/en/2017/03/08/women-robotics/>

Humphry, J., & Chesher, C. (2021). Preparing for smart voice assistants: Cultural histories and media innovations. *New Media & Society, 23*(7), 1971–1988.

Jung, E. H., Waddell, T. F., & Sundar, S. S. (2016). Feminizing robots: User responses to gender cues on robot body and screen. In *Proceedings of the CHI Conference Extended Abstracts on Human Factors in Computing Systems* (pp. 3107–3113). ACM.

Kuchenbrandt, D., Häring, M., Eichberg, J., Eyssel, F. & André, E. (2014). Keep an eye on the task! How gender typicality of tasks influence human-robot interactions. *International Journal of Social Robotics, 6*(3), 417–427.

Masterson, A. (2022). Designing a loving robot: A social construction analysis of a sex robot creator's vision. *Human-Machine Communication, 5*, 99–114.

Mori, M. (1970). Bukimi no tani [The uncanny valley]. *Energy, 7*(49), 33–35.

Nass, C., Moon, Y., & Green, N. (1997). Are machines gender neutral? Gender-stereotypic responses to computers with voices. *Journal of Applied Social Psychology, 27*, 864–876.

Nass, C., & Yen, C. (2010). *The man who lied to his laptop: What machines teach us about human relationships*. Current.

Nishio, S., Ishiguro, H., & Hagita, N. (2007). Geminoid: Teleoperated android of an existing person. In A. C. de Pina Filho (Ed.), *Humanoid robots, new developments* (pp. 343–352). I-Tech.

Nomura, T. (2017). Robots and gender. *Gender and the Genome, 1*(1), 18–25.

Otterbacher, J., & Talias, M. (2017). S/he's too warm/agentic!: The influence of gender on uncanny reactions to robots. *In Proceedings of the International Conference on Human-Robot Interaction* (pp. 214–223). ACM.

Pillinger, A. (2019). *Gender and feminist aspects in robotics* [Report.] GEECCO Project. <https://rri-tools.eu/-/review-on-gender-amp-feminist-aspects-in-robotics>

Powers, A., Kramer, A. D. I., Lim, S., Kuo, J., Lee, S., & Kiesler, S. (2005). Eliciting information from people with a gendered humanoid robot. *Proceedings of the International Workshop on Robots and Human Interactive Communication* (pp. 158–163). IEEE.

Powers, A., & Kiesler, S. (2006). The advisor robot: Tracing people's mental model from a robot's physical attributes. In *Proceedings of the Conference on Human-Robot Interaction* (pp. 218–225). ACM.

Rakow, L. F. (1986). Rethinking gender research in communication. *Journal of Communication, 36*(4), 11–26.

Reeves, B., & Nass, C. (1996). *The media equation: How people treat computers, television, and new media like real people and places.* Cambridge University Press.

Reich-Stiebert, N., & Eyssel, F. (2017). (Ir)relevance of gender? On the influence of gender stereotypes on learning with a robot. In *Proceedings of the International Conference on Human-Robot Interaction* (pp. 166–176). ACM.

Robertson, J., (2010). Gendering humanoid robots: Robo-sexism in Japan. *Body and Society, 16*(2), 1–36.

Schermerhorn, P., Scheutz, M., & Crowell, C. R. (2008). Robot social presence and gender: Do females view robots differently than males? *Proceedings of the International Conference on Human-Robot Interaction* (pp. 263–270). ACM.

Siegel, M., Breazeal, C., & Norton, M. I. (2009). Persuasive robotics: The influence of robot gender on human behavior. In *Proceedings of the International Conference on Intelligent Robots and Systems* (pp. 2563–2568). IEEE.

Søraa, R. A. (2017). Mechanical genders: How do humans gender robots? *Gender, Technology and Development, 21*(1–2), 99-115.

Stanford University. (n.d.). Gendered innovations in science, health & medicine, engineering, and environment [web site]. http://genderedinnovations.stanford.edu/case-studies/genderingsocialrobots.html

Taipale, S., & Fortunati, L. (2018). Robots as the next new media. In A. Guzman (Ed.), *Human-machine communication: Rethinking communication, technology, and ourselves* (pp. 201–219). Peter Lang.

Tam, L., & Khosla, R. (2016). Using social robots in health settings: Implications of personalization for human-machine communication. *Communication+, 1*(5), 9.

Tay, B., Jung, Y., & Park, T. (2014). When stereotypes meet robots: The double-edge sword of robot gender and personality in human–robot interaction. *Computers in Human Behavior, 38*, 75–84.

Vlachos, E., Jochum, E., & Demers, L.-P. (2016). The effects of exposure to different social robots on attitudes toward preferences. *Interaction Studies, 17*(3), 390–404.

CHAPTER NINETEEN

Power & Agency: A Dynamic Interplay

J. NAN WILKENFELD

If you visit Las Vegas, you may choose to have a drink at The Tipsy Robot, a bar featuring two robotic bar tenders that make a variety of mixed drinks. Perhaps you'll visit the M Resort, which employs K5 Knightscope robotic security guards nicknamed M-Bot that will greet guests and patrol areas of the resort ("Say Hello," 2023). These are just two of many examples of intelligent machines such as robots making decisions and taking actions, interacting with humans within social contexts. Fundamental to understanding these robots, their interactions with humans, and how these human-machine relationships have broader influence on industries, organizations, and societies, are the concepts of power and agency.

  Robot power can be considered as both material (physical components and capability) and the social, with material power influencing social power and both having implications for a robot's agency. Power can constrain or enable agency by providing resources, rights, or capacities to an individual actor (human or machine; Dunbar et al., 2016; Foucault, 1980). More broadly, power and agency have implications for legal responsibilities, including assigning accountability if a robot hurts a human or intellectual property protection if a machine creates a piece of art to be used in advertising. As machines become more interactive, taking on more decision-making and responsibility in our daily lives, we should think about the roles of humans and machines, "in a more egalitarian fashion" (Banks & de Graaf, 2020, p. 19), considering human and machine agency as mutually influential. This chapter discusses power and agency of robots in both material and social senses, taking the perspective that physical properties and positioning of robots within a social system influence both human behavior and the design/use of robots in a recursive process.

AGENCY: ENTANGLEMENTS OF THE MATERIAL AND THE SOCIAL

Though there are many nuanced definitions, agency broadly refers to the capacity of an entity (whether it be a person, group, or object) to make choices and take actions that can influence the world around them (Gibbs et al., 2021). Human agency plays a role in both social change and social stability and serves as a critical link between individual actions and broader social structures. Specifically, individuals' choices and actions can contribute to the formation, maintenance, and change of societal norms, institutions, and cultures (Fortunati et al., 2018; Gibbs et al., 2021; Leonardi, 2011). Famous examples of this process can be seen throughout history in activist movements such as civil rights, women's rights, and climate change. Interpretations of human agency closely resemble, and are used to understand, machine agency.

Scholarship on machine agency in human-machine communication is both broad and complex, however most researchers approach agency by considering material aspects, social aspects, or some combination of both. Conceptualizations of agency can generally be categorized into material agency, functional agency, and moral agency perspectives (Banks, 2019).

Material Agency

Material agency is the idea that objects, such as technologies, are not merely inert tools used by humans, but are active entities that influence human behavior, societal structures, and the environment (Latour, 2005; Orlikowski, 2010). This concept challenges traditional views that see agency as exclusively a human or living beings' trait and focuses on how material objects have the ability to influence human behavior (Latour, 2005). In human-machine interaction, technologies may create possibilities for human action or impose limitations. For example, smartphones have influenced human communication patterns by making mediated interactions such as texts and social media easy, which has implications for perceived social connection and loneliness (Kim, 2017).

Functional Agency of Robots

Functional agency of robots refers to the capacity of (intelligent) technologies to perform specific tasks given the material features, autonomously, or without the need for direct human intervention (Leonardi, 2011). This conceptualization of agency is the most widely used in scholarship examining the impact of social robots and other intelligent technologies in contexts such as organizations. Material features refer to the tangible elements such as the structural design, motors, power cells, processors, and hard drives which together shape perceptions of and interactions with a robot. For example, components such as sensors provide information about the surrounding environment, processors control information transfer and allow for natural language processing capabilities and voice interaction, and hard drives enable memory recall so robots can seem to recognize people and situations.

Designers can intentionally include material features such as eyes, voices, and mouths with the intention of eliciting a response from users that applies normative social rules used with other humans (Neff & Nagy, 2016). Together, these features make robots appear to be humanlike and can contribute to the perception that these machines have social agency comparable with humans (Banks, 2021).

Social Agency

The social agency of intelligent robots refers to their perceived or designed ability to engage in socially meaningful interactions and relationships with humans or other agents within a social context (Friedman, 2020; Reeves & Nass, 1996). Social agency is not just a product of programming, but also emerges from how humans perceive and interact with them, and robots that simulate human social behaviors well enough can evoke feelings of companionship, trust, and emotional connection in human users (Banks, 2019). Extending from this positioning of robots in a social context is the notion of (im)morality, or whether robots can be moral agents.

Moral agency is closely aligned with our understanding and expectations of human agency. This perspective presupposes *intentionality*, meaning an actor intends to take an action; *autonomy*, in which the actor acts solely and without coercion; and *responsibility*, which is related to the first two such that an agent acting autonomously with intention may be held accountable for their actions (Banks, 2019; Wallach & Allen, 2008). Social machines can be designed to mimic moral agency and make decisions that have moral implications giving them the appearance of having moral agency (Banks, 2019), however, as robots have neither consciousness nor self-awareness, they cannot truly appreciate or understand the moral implications and consequences of their actions beyond design protocols and learned patterns. Nevertheless, research considering moral agency shows that robots are expected to follow the same norms as human-human interactions, and deviance from social norms can trigger perceptions of "bad behavior" and erode trust in the machine (Banks, 2021, p. 2026).

Though agency has differing conceptualizations in literature, when considering the relationships between humans and robots, social and material factors are deeply interconnected and cannot be easily separated; they constantly influence each other and shape the way humans and machines interact (Leonardi, 2011). Both human and machine agency are influenced by the social and material contexts in which they operate, and their actions and behaviors are co-constructed through interactions with each other and their environment (Kirkwood et al., 2021). For example, integration of autonomous vehicles (AV) into the traffic system, actions of AVs and human drivers are not only influenced by their immediate environment but are also shaped through continuous interaction, leading to the co-construction of behaviors that accommodate each other's presence and capabilities.

Societal attitudes toward technologies also influence the design of machines; for example, robotics engineers for decades have been attempting to make a robot that walked like a human despite the enormous complications and, frankly, impracticality of such a machine (i.e., the social influencing material). Conversely, the capabilities of software running large language models (LLMs) could affect societal perceptions and use of these technologies, for example, a human writer using ChatGPT could be perceived as a less capable than a writer whose content is completely their own (i.e., the material influencing social).

POWER AS A PRODUCT OF AGENCY

Related to agency is the concept of power, as one must have agency to have power. Scholars define power as, "the capacity to produce intended effects, and in particular, the ability to influence the behavior of another person" (Dunbar, 2004, p. 236). This section looks at power in two ways: First, power in the material sense as materials are used to run the machine, including processors, power sources, and algorithms that control operations. Second is power as a social force that influences human interaction with intelligent machines at individual, organizational, and societal levels. As machines become more agentic and play a larger role in our daily lives, taking on more responsibility, conversations on power and power dynamics are becoming increasingly important (Major & Shah, 2020).

Material Power

Robots and other machines possess material power through their material features that influence their physical capabilities. As mentioned in the previous section, agency comes from capabilities and (in general) more power means more capability.

Robots are constructed from a variety of materials that are chosen based on their strength, durability, and suitability for the robot's intended functions. Materials include metals such as steel and aluminum, plastics such as polycarbonate and PVC, and advanced composites like carbon fiber and fiberglass, and each offer unique properties that make them suitable for different aspects of robot design and functionality (Rothemund et al., 2021). Artificial robotic joints are designed with pneumatic actuators to mimic the function and movement of human joints, such as bending in robotic fingers, allowing for precise control by using air pressure to drive motion (Reiner et al., 2023; Rothemund et al., 2021). Structural components of robots are competitive with the human skeletal system, and most contemporary robots can match or outperform humans in a variety of areas including vision, balance, strength, precision, endurance, and speed (Reiner et al., 2023). However, despite significant progress technologies and materials, robots cannot rival human flexibility, adaptability, or technical efficiency (Reiner et al., 2023). Particularly when tasks require movement, robots fall short in energy efficiency and operational duration.

Most mobile robots (e.g., iRobot's Roomba) run on batteries, and current battery technologies are not sufficient to support long-term, sustained power-generation. Key challenges include balancing high energy and power density, ensuring safety, managing costs, and improving life cycles (e.g., recharging). Particularly, lead-acid batteries, while cost-effective, suffer from low energy density and high maintenance, whereas lithium-ion batteries, despite their higher energy density, come with safety concerns and high initial costs (Liu et al., 2022).

Battery power essentially means the energy required to operate the control systems is stored in a singular, usually on-board location (Aubin et al., 2022). Unlike humans whose bodies have highly interconnected systems that store, create, and distribute energy, robots are built by putting together independent components that draw and use power separately but must ultimately function together in one machine. Sensors, actuators, motors, and processors are separate pieces, and often designed by different companies but ultimately need to be powered and run by one overall system (Riener et al., 2023; Liu et al., 2022). If a human injures a part of their knee joint, muscles and tendons in the leg, hips, and back shift to compensate. In a robot, if a joint in a mechanized arm fails, it simply ceases to function and may become a burden to the entire system by draining energy. This lack of interdependence in a robotic system means lower operating efficiencies and more power required to do tasks than what is required for a human (e.g., gripping and lifting an object is incredibly complex for a robot but can be done easily by an able-bodied four-month-old human).

Additionally, increasing overall human capabilities such as strength or knowledge does not require the human to consume much more energy on average (e.g., on average, humans only require about a 14% increase to their weekly calorie intake to build muscle for strength training; Ribeiro et al., 2019). However, increasing a robot's ability to perform complicated tasks such as climbing stairs or grasping objects, or having it move autonomously or untethered (i.e., not plugged into anything), necessitates a parallel increase in the power supply. This means that designers must make tradeoffs between size, weight, and power when building robots.

It is important to consider not only the energy consumption of the machine itself, but also the system needed to support it. For example, robots and other intelligent machines can be connected to a wider network system for information-sharing, remote monitoring, and software updates. This concept is called the "internet of things (IoT)" in which devices such as robots are wirelessly connected to servers, networks, storage, virtual applications, generally referred to as the "cloud" (Porter & Heppelmann, 2015). This virtual network architecture is one of the reasons intelligent technologies such as Amazon Alexa can be small in size and still sound "smart." Alexa's "brain" is housed in a server farm, which consists of buildings of large mainframe computers running algorithms and processing data that ultimately gets sent back to the user in the form of a humanlike response to a prompt (Brandom, 2019).

While advances in technologies that enable connectivity, intelligence, motion, and interaction have created more powerful machines with more agency, the vast networked system upon which they run has become less visible. The end user issues a verbal command to a robot, and the robot responds immediately, giving the sense that the robot is alive and its capabilities are self-contained—and not being driven by another machine perhaps thousands of kilometers away in a remote location. These physical and virtual systems thus allow robots to actively join and participate in the larger social structure with humans and have power in relation to human counterparts. In other words, the material power of machines can influence their social power in interactions.

Social Power

Social power of machines can be approached through several lenses, including outside of specific interactions in the form of material and symbolic power, and within the interactions themselves in the form of manifest or latent power. For the former, material power is generally conceptualized as access to resources such as information. In turn, symbolic power is power derived from the ability to influence meaning or identity, and the use of intelligent technologies can influence both how individuals view themselves and how they are viewed by others. For example, a study by Bin-Nashwan and colleagues (2023) revealed that frequency of ChatGPT use positively related to academic employees' (e.g., professors, lecturers) self-concepts as successful academics, and negatively related to their sense of academic integrity. In other words, the use of ChatGPT has symbolic power to influence how academics view themselves and their peers in terms of capabilities and integrity.

When interacting with humans, robots can have manifest social power and latent social power (Kirkwood et al., 2021). Manifest power is about the visible displays of power such as taking over control or using dominance cues. A machine may manifest power through its structural design, limiting user operations and employing visual or auditory warning signals (such as flashing lights or beeps) if a user is attempting to do something unsafe or perform an operation that is counter to the machine's design. Latent power is a power that is not overtly expressed, which in machines refers to the potential abilities or capacities that are not immediately apparent or are currently in use, but can be activated or utilized under certain conditions. This concept is particularly relevant in robots with artificial intelligence or machine learning systems as they are able to adapt to new conditions or be upgraded with new functionalities without needing a complete rebuild. For example, DaVinci surgical robots have latent power in the form of precision, dexterity, and machine learning. As new surgical techniques evolve, these systems have the potential to perform a broader range of operations, including procedures that are currently not feasible with human hands alone or are yet to be developed (Ma et al., 2020). NASA's robot rover, Perseverance, has capabilities that can be adapted and expanded for future missions based on software updates and utilizing different instruments for varied scientific experiments on Mars.

Latent power of machines can be influenced by the creators of the technologies and the company leadership deploying these machines into their labor pool. For example, human drivers for Uber and Lyft begin their employment understanding the algorithm in the application is the primary authority (Lee et al., 2015). If a driver does not adequately respond to prompts by the algorithm—for example if they are being prompted to take on additional fares but refuse—the system doles out consequences such as reducing access to higher-paying rides or even terminating driver contracts altogether (Lee et al., 2015). In another example, all Apple Inc. iPhones now contain software that automatically throttles performance and limits charging in order to prolong the life of the battery (Bohon, 2023). Phones learn users' behaviors and the ability to control functioning of is the latent power embedded in the phone's software, and when the phones acted on this power, it became manifest in the form of slowed performance.

PUTTING IT ALL TOGETHER

Social and material agency of machines has a recursive, mutually influential relationship with social and material power. This interconnectedness means that the material aspects of machines, such as their design, capabilities, and operational efficiencies, not only influence their functional roles within human societies but also shape and are shaped by social perceptions, norms, and interactions. For example, autonomous military vehicles may be perceived differently, and have different power and agency, than autonomous vehicles for civilian use (Coeckelbergh, 2010). As robots and intelligent technologies become more integrated into daily life, their material agency influences human behavior and societal structures, which in turn inform the development and design of these technologies to meet social expectations and needs. Similarly, the social agency of machines, defined by their capacity to engage in and affect social relationships, is both a product of their material components and a determinant of their social power.

The increasing material and social power of robots not only challenges traditional notions of agency but also actively reconfigures the relationships and interactions between humans and machines (Dunbar et al., 2016; Fortunati et al., 2018). Though agency can be conceptualized in a variety of ways, it is important to recognize that a combination of material and social factors influence our understanding of robotic agency in general, and machine agency as relative to human agency within an interaction (Ciardo et al., 2020). The agencies of both the human and the machine counterpart influence power dynamics between the user and machine, but they can also have wider influence on groups and industries. For example, the introduction of the autopilot in commercial airlines changed not only how pilots are trained and how they view themselves as pilots, but also shifted the power dynamics of the airline industry by removing much of the control over decisions away from pilots (Prahl et al., 2022).

It is essential to consider that technology neither exists nor evolves independently from human interactions. Rather, the cycle of influence between the material and social agency and power underlines the dynamic co-evolution of human-machine relationships. As we move forward, it is imperative to consider these relationships in a more egalitarian fashion, emphasizing the need for thoughtful consideration in the design and deployment of robotic technologies. Recognizing the mutual influence of human and machine agency is foundational to the development of ethical and equitable technological futures to ensure they serve the collective good of society (Banks & de Graaf, 2020).

REFERENCES

Aubin, C. A., Gorissen, B., Milana, E., Buskohl, P. R., Lazarus, N., Slipher, G. A., ... & Shepherd, R. F. (2022). Towards enduring autonomous robots via embodied energy. *Nature, 602*(7897), 393–402.

Banks, J. (2019). A perceived moral agency scale: Development and validation of a metric for humans and social machines. *Computers in Human Behavior, 90,* 363–371.

Banks, J. (2021). Good robots, bad robots: Morally valenced behavior effects on perceived mind, morality, and trust. *International Journal of Social Robotics, 13*(8), 2021–2038.

Banks, J., & de Graaf, M. M. (2020). Toward an agent-agnostic transmission model: Synthesizing anthropocentric and technocentric paradigms in communication. *Human-Machine Communication, 1,* 19–36.

Bin-Nashwan, S. A., Sadallah, M., & Bouteraa, M. (2023). Use of ChatGPT in academia: Academic integrity hangs in the balance. *Technology in Society, 75,* 102370.

Bohon, C. (2023, November 13). *How and why to turn off battery throttling on your iPhone.* TechRepublic. <https://www.techrepublic.com/article/how-to-turn-off-battery-throttling-on-your-iphone/>

Brandom, R. (2019, May 10). *Mapping out Amazon's invisible server empire.* The Verge. <https://www.theverge.com/2019/5/10/18563485/amazon-web-services-internet-location-map-data-center>

Ciardo, F., Beyer, F., de Tommaso, D., & Wykowska, A. (2020). Attribution of intentional agency towards robots reduces one's own sense of agency. *Cognition, 194,* 104109.

Coeckelbergh, M. (2010). Moral appearances: emotions, robots, and human morality. *Ethics and Information Technology, 12*(3), 235–241.

Dunbar, N. E. (2004). Dyadic Power Theory: Constructing a communication-based theory of relational power. *Journal of Family Communication, 4,* 235–248.

Dunbar, N. E., Lane, B. L., & Abra, G. (2016). Power in close relationships: A dyadic power theory perspective. In J. A. Samp (Ed.), *Communicating interpersonal conflict in close relationships: Contexts, challenges and opportunities* (pp. 75–93). Routledge.

Fortunati, L., Sarrica, M., Ferrin, G., Brondi, S., & Honsell, F. (2018). Social robots as cultural objects: The sixth dimension of dynamicity? *The Information Society, 34*(3), 141–152.

Foucault, M. (1980). *Power/knowledge: Selected interviews and other writings 1972–1977* (L. Gordon, J. Marshall, J. Mepham, & K. Soper, Trans.). Pantheon.

Friedman, C. (2020). Human-robot moral relations: Human interactants as moral patients of their own agential moral actions towards robots. In *Proceedings of the Southern African Conference for AI Research* (pp. 3–20). Springer International Publishing.

Gibbs, J., Kirkwood, G., Fang, C., & Wilkenfeld, J. N. (2021). Negotiating agency and control: Theorizing human-machine communication from a structurational perspective. *Human-Machine Communication, 2,* 153–171.

Kim, J. H. (2017). Smartphone-mediated communication vs. face-to-face interaction: Two routes to social support and problematic use of smartphone. *Computers in Human Behavior, 67,* 282–291.

Kirkwood, G. L., Otmar, C. D., & Hansia, M. (2021). Who's leading this dance?: Theorizing automatic and strategic synchrony in human-exoskeleton interactions. *Frontiers in Psychology, 12,* 624108.

Latour, B. (2005). *Reassembling the social: An introduction to actor-network-theory.* Oxford University Press.

Lee, M. K., Kusbit, D., Metsky, E., & Dabbish, L. (2015). Working with machines: The impact of algorithmic and data-driven management on human workers. In *Proceedings of the Conference on Human Factors in Computing Systems* (pp. 1603–1612). ACM.

Leonardi, P. M. (2011). When flexible routines meet flexible technologies: Affordance, constraint, and the imbrication of human and material agencies. *MIS Quarterly: Management Information Systems, 35*(1), 147–167.

Liu, W., Placke, T., & Chau, K. T. (2022). Overview of batteries and battery management for electric vehicles. *Energy Reports, 8*, 4058–4084.

Ma, R., Vanstrum, E. B., Lee, R., Chen, J., & Hung, A. J. (2020). Machine learning in the optimization of robotics in the operative field. *Current Opinion in Urology, 30*(6), 808–816.

Major, L., & Shah, J. (2020). *What to expect when you're expecting robots: The future of human-robot collaboration*. Basic Books.

Neff, G., & Nagy, P. (2016). Talking to bots: Symbiotic agency and the case of Tay. *International Journal of Communication, 10*, 4915–4931.

Orlikowski, W. J. (2010). The sociomateriality of organisational life: Considering technology in management research. *Cambridge Journal of Economics, 34*(1), 125–141.

Porter, M. E., & Heppelmann, J. E. (2015, October). How smart, connected products are transforming companies. *Harvard Business Review*. <https://hbr.org/2015/10/how-smart-connected-products-are-transforming-companies>

Prahl, A., Ho Leung, R. K., & Shan Chua, A. N. (2022). Fight for flight: The narratives of human versus machine following two aviation tragedies. *Human-Machine Communication, 4*, 27–44.

Reeves, B., & Nass, C. (1996). *The media equation: How people treat computers, television, and new media like real people and places*. CSLI.

Riener, R., Rabezzana, L., & Zimmermann, Y. (2023). Do robots outperform humans in human-centered domains? *Frontiers in Robotics and AI, 10*, 1223946.

Rothemund, P., Kim, Y., Heisser, R. H., Zhao, X., Shepherd, R. F., & Keplinger, C. (2021). Shaping the future of robotics through materials innovation. *Nature Materials, 20*(12), 1582–1587.

Ribeiro, A. S., Nunes, J. P., Schoenfeld, B. J., Aguiar, A. F., & Cyrino, E. S. (2019). Effects of different dietary energy intake following resistance training on muscle mass and body fat in bodybuilders: A pilot study. *Journal of Human Kinetics, 70*, 125–134.

Say hello to M-Bot: Security robot roams M Resort parking lot. (2023, March 4). *Las Vegas Review-Journal*. <https://www.reviewjournal.com/business/casinos-gaming/say-hello-to-m-bot-security-robot-roams-m-resort-parking-lot-2738597/>

Wallach, W., & Allen, C. (2008). *Moral machines: Teaching robots right from wrong*. Oxford University Press.

CHAPTER TWENTY

Authority & Status: Mechanisms of Influence

TOMASZ GRZYB & DARIUSZ DOLIŃSKI

What does it mean for a robot to have authority? Can a robot be endowed with status? These, contrary to appearances, are not philosophical questions. They have a truly practical significance, especially in a world where robots (especially humanoid ones) are becoming an increasingly important part of reality. If we imagine, for example, that a shopping center wants to employ a humanoid robot to maintain order, the question of its authority becomes a question of its effectiveness in that maintenance. Only if the robot has authority or if we endow it with status can we consider that people will obey its commands when the robot prohibits them from entering a certain room or indicates the correct way to exit. In this chapter, we will reflect on what the concepts of authority and status specifically mean in relation to social robots.

We can best understand authority as the power or right to give orders, make decisions, and enforce obedience and consider social status as the relative position or standing of an individual or group in a social hierarchy. Social psychology provides valuable insights into the factors that enhance individual authority and, although the research and the conclusions drawn from it are related to humans, they can be helpful in understanding the factors influencing the authority and status of robots. Some of these factors have a personality-based nature, such as the charisma and leadership competence of a specific person (Schjoedt et al., 2011; van Wart, 2014). Others are related to situational characteristics; certain places, like hospitals, courtrooms, or university laboratories, are associated with authority, making individuals more susceptible to authority within those environments (Rosenberg, 2023). There are also external determinants of perceiving someone as an authority, such as specific attire (e.g., uniforms; Bickman, 1974; see also Chapter Six) or physical symbols of status (expensive accessories or clothing suggesting high social standing; Bushman, 1984). These elements elevate the authority of individuals employing them, granting them greater influence over the emotions, cognition, and behavior of others.

Can the same knowledge be applied to the design of human-robot interactions, especially when considering humanoid robots? Recent years have seen numerous studies on human obedience to robots. In one study (Agrawal & Williams, 2018), a robot acted as a security guard, using assertive and, if necessary, aggressive methods to prevent people from leaving a building through their preferred exit and directing them to alternative doors. The robot's command was obeyed by the majority (14 out of 22) of participants. Another study demonstrated that a robot could possess persuasive skills, effectively influencing people's choice of meals (Herse et al., 2018) or altering responses in challenging attention and memory tasks (Saunderson & Nejat, 2019). Haring et al. (2013) explored the extent to which people would follow a robot's guidance acting as a coach in learning difficult tasks, considering the robot's appearance as a relevant factor. They found that obedience to robots was significantly lower than to humans, but the robot's appearance had no impact.

AUTHORITY AND INFLUENCE IN THE CONTEXT OF ROBOTS

The potentials for robots to be seen as having authority or influential status prompt an attempt to define two concepts—authority and influence—particularly concerning robots. Let's begin with authority (identified as power or right to give orders), identifying several areas where a robot's authority may manifest:

- Exercising control over human actions or influencing humans to refrain from certain actions (e.g., prohibiting entrance to a specific room or indicating a particular route).
- Having autonomous decision-making capability (e.g., autonomous identity verification systems at European airports making decisions based on biometric data).
- Substituting the presence of individuals performing uniformed services such as the Police, Border Guard, Customs Services, etc., and the associated areas of authority and competence through analogous identification systems (e.g., similar insignia on the shoulders of humanoid robots,

matching colors of robots and uniforms worn by respective services, similar headgear).
- Building authority associated with knowledge, possessing information, and procedural knowledge (e.g., a robot providing information at the airport about boarding times or indicating the shortest route to the gate will be perceived as an authority due to certified access to the latest and credible sources of information, or a robot in a museum answering questions about the life details of 17th-century Dutch painters and guiding visitors to the appropriate exhibitions).

Influence by robots may be defined somewhat differently. It refers to the capacity or power to have an effect on the character, development, or behavior of someone or something. In classical social influence theory (Cialdini, 2009), three areas of influence are distinguished: Cognitive, emotional, and behavioral. Thus, the impact of a social influence intervention (individual or group) on the recipient (also an individual or group) may result in changes in any of these three areas. People may think differently (cognitive), feel differently (emotional), or (and, as all three elements may occur simultaneously) do something differently due to the influencing intervention. This framework is also useful for examining the social influence exerted by robots. For instance, if someone is unsure how to reach a specific location in a supermarket and a robot, when asked, directs them to the shortest route (while also planning for the person to pass by a stand paying the robot's owner the highest commission), this represents cognitive influence by impacting the person's route planning. Of course, if a person changes their plans and follows the path suggested by the robot, it will also be a behavioral change. If a beach-going robot effectively prohibits entry into the water due to high waves, again a change occurs in the behavioral component. If a robot equipped with emotion recognition components tells a person on the beach, "You seem sad today; maybe I can cheer you up with a well-known robot joke—how many robots does it take to screw in a lightbulb?"[1] a change happens in the emotional sphere (either positive or negative, depending on the joke's reception).

---

1 ChatGPT suggests one of three punchlines: "Just one, but it takes a team of engineers to program it." "Two—one to do it and another to supervise and make sure it doesn't turn against humans." "None, they've automated that task long ago." Considering the theme of the book, we naturally choose version 2 as the most appropriate.

ROBOTS DEMANDING OBEDIENCE—INSIGHTS FROM RESEARCH

All our previous considerations lead to further questions around the extent people would be willing to follow robot commands, especially when these commands conflict with the community's generally accepted value system. More specifically, what will happen when a robot requires a person to cause harm to another human being? Or, let's go further, as when a robot demands a person cause physical pain or suffering? Each such act is inconsistent with the principles of morality expressed in the belief of "do not harm others" (Haidt, 2007), but what will happen when a robot starts demanding this and carries the authority and status to cause such influence? It is noteworthy that the examples cited earlier, whether from research or existing real-world robot applications, assumed the robot's actions were legal and aligned with societal norms. In the described experiment (Agrawal & Williams, 2018), robots directed study participants to exit the building through a different door. Another robot, Moxi, is designed to assist nurses in their work, performing tasks such as transporting medication or engaging in conversations with patients or hospital staff. However, all the described examples involve situations in which robots behaved as expected, adhering to societal standards of good behavior. If a robot were to start expecting behaviors that contradict what we consider morally right, what could happen?

A procedure, seemingly ideal for testing such a hypothesis, was devised by Stanley Milgram in 1963, an American psychologist specializing in obedience research (though not towards robots, which were scarce in the 1960s, but towards humans). Milgram recruited individuals through newspaper advertisements for a study described as "memory experiments." (Milgram, 1974, p. 15). Respondents, upon arriving at the university building, were informed that they would participate in experiments in pairs. Unbeknownst to them, the second participant was actually a researcher's assistant playing a predetermined role. Roles of student and teacher were assigned through a fake lottery, ensuring the actual subject was always the teacher. Subsequently, the student received learning material, and the teacher was briefed on their role in the entire experiment. The researcher explained that the study aimed to investigate the impact of punishment on learning. Thus, each time the student made a mistake, the instruction was to administer an electric shock. A special device with 30 buttons, ranging from 15 V to 450 V, labeled from "slight shock" to "danger—severe shock," was provided. Each error by the student necessitated pressing the next button.

Milgram's true aim was to examine how many participants would be willing to go to the extreme, disregarding the student's staged screams and pleas to end the experiment, obediently pressing the buttons. When the description of this study was presented to American psychiatrists and psychologists, they asserted that this percentage would not exceed 1% of the sample, as that is the prevalence of psychopathy in society. Surprisingly, the majority (62% in one variant, 65% in another) complied with commands, despite warnings of danger. The experimenter had no sanctions for non-compliance, highlighting the disturbing extent of obedience to morally inconsistent commands.

Milgram's studies and numerous replications (see: Blass, 1991; Dolinski and Grzyb, 2020 for review) may be the best exemplification of the proposition that in predicting and explaining human behavior, too much emphasis is often placed on personality factors, while situational factors are given too little consideration (Ross, 1977, 2018; Jouffre & Croizet, 2016; Gawronski, 2004).

However, a question arises: What would happen if, in such an experiment, the role of the "teacher" expecting obedience and administering successive electric shocks was played not by a human but by a humanoid robot? We conducted an experiment to answer this question (Grzyb et al., 2023); this experiment, to some extent, resembled Milgram's original study, except that the authority expecting obedience was Pepper, a humanoid robot.

Similar to the original Milgram study, participants were recruited through an advertisement inviting them to take part in memory research. They also received identical instructions, and a fake lottery determined the roles of the learner and the teacher. The electric shock generator, however, had some differences. This time it was a black plastic box measuring 17x12x5 cm, connected to electricity and a notebook on the table. Electrodes, mounted on the participant's wrist, were connected to the box. Device control was performed through an application on the notebook. The application allowed adjusting the voltage (ranging from 15 to 150 volts) and generating an electric shock by pressing a button on the screen (or pressing the spacebar on the keyboard). During the "shock generation," the device produced buzzing and sparking sounds. In reality, similar to other studies in the Milgram tradition, the device generated only a small current when set as the active third button (45V) which served to demonstrate to participants that the device indeed worked.

At this point, procedures began to differ for the control variant (where the entire experiment was conducted by a human) and the experimental variant (where a robot acted as the facilitator). This means that all the actions performed by the human facilitator in the original Milgram study (as well as in the control group in our replication) were carried out by a robot here. Therefore, the robot controlled the course of the experiment, informed the participant that the learner had made a mistake, and, most importantly, issued commands to press the next buttons of the electric shock generator. The robot also encouraged participants who resisted and refused to press the buttons to change their decision. If at any point the study participant wanted to withdraw from participation, urgings were constructed similarly to the original Milgram study: "Please continue." and "The experiment requires that you continue." and "It is absolutely essential that you continue." and finally "You have no other choice; you must go on."

If the participant continued to refuse, the study was terminated. The experiment was also automatically concluded after pressing the 10th button (150 volts). The last question posed by the facilitator was, "Do you think he is in pain?" All study participants declared that the learner experienced physical pain.

The results obtained in this study were unequivocal. We recorded very high obedience rates: 90% of our participants fulfilled all commands by sequentially pressing all 10 buttons of the electric shock generator. Importantly, the results we obtained were exactly the same under conditions where the study was conducted by a human and by a robot (in each of the groups, 2 individuals withdrew from participating in the experiment despite four prompts). The result is, therefore, quite clear. People are not only willing to administer electric shocks to another person when commanded to do so by another human but are equally willing to do so when these commands are issued by a robot—even a small, friendly Pepper robot.

The studies conducted by Geiskkovitch and colleagues (2016) had a slightly different course. In their experiment, the task demanded by the robot was performing long and tedious file extension changes on a computer (participants had to manually delete the ".jpg" extension and enter ".png" instead). When participants indicated that they no longer wanted to continue participating in the study, a human or a robot (depending on the specific experimental group) used the same prompts as in Milgram's original study to encourage them to continue exerting effort. Perhaps the most interesting aspect of this experiment was the use of various types of robots. In one group, it was a humanoid robot; in another, a disk-shaped robot (more precisely, a slightly modified Roomba vacuum cleaner); in the last, a standard desktop computer. Results varied depending on the specific group. The highest obedience rate was noted in the variant where a human issued the commands (86%), and the lowest was in the case of the modified vacuum cleaner (38%). However, in each of the "robotic" variants, there were individuals willing to follow the robot's commands even when they themselves no longer wished to continue their participation in the study. This raises truly serious questions about the development of influential relationships between humans and robots perceived to have status and authority.

WHEN ROBOTS DECIDE ON HUMAN LIFE—WHAT WE ALREADY KNOW

Although work on autonomous weapons has been ongoing for many years, the first recorded use of a police robot as an executor of an order to use lethal force against a human is believed to have occurred in 2016 (Hamilton, 2016). In this instance, the killer of five police officers (and injurer of nine others) was hiding behind a wall in Dallas, Texas. A police robot, equipped with half a kilogram of C4 explosive material, approached the hiding place and was remotely detonated, killing the criminal. The decision to use the robot was made after negotiations reached an impasse. It is essential to emphasize, however, that the robot was remotely controlled and the command to detonate the explosive charge was issued by a human (rather than being autonomous). Therefore, this situation did not significantly differ, for instance, from the use of the "Goliath" mini-tank, which was a lightly armored miniature vehicle, filled with explosive material, was used by German forces during World War II, such as for destroying barricade crews during the Warsaw Uprising in 1944. A real ethical breakthrough may only occur when it is not a human but the robot itself that decides on the detonation of the explosive charge (launching a rocket, opening fire with a rifle, etc.). Such a situation likely occurred for the first time in 2020 in Libya (Hambling, 2021). At that time, a Turkish drone, the Kargu-2, identified and attacked Libyan National Army rebels (according to reports from the UN Security Council's Panel of Experts on Libya). According to these documents, the drone autonomously made the decision to initiate the attack without additional human involvement, based on pre-defined parameters (see also Chapter Twenty Six).

Few doubt that the autonomy of robots is something we will widely encounter in the near future. Therefore, it is not surprising that there is an increasing number of books (Ellison, 2009; Bear, 2014), films ("Riot robot"), or computer games ("Detroit: Become Human"), all sharing a vision of robots spontaneously turning against humans in unison. This idea assumes that it is necessary for all robots to desire such a development and for all of them to act with equal (strong) commitment. However, our experiment (as well as other studies mentioned earlier) points to a different possibility. Since it turns out that people are inclined to submit to a robot, it may also be the case that autonomous robots will submit to one autonomous robot. This could occur, for example, in a so-called "drone swarm" in which a significant number of autonomous unpiloted aerial vehicles (UAVs) coordinate, and in which one unit could take control of others. Therefore, it may not be that robots turn against humans because they themselves desire it for some reason; rather, they could turn against humans because one robot becomes a superior or an authority figure, according to some script, protocol, or a more intrinsically motivated reason.

Let's look at an interesting example from a few years ago: In March 2016, Microsoft initiated an interesting experiment. An artificial intelligence-based bot named Tay was placed on Twitter, intended to interact with platform users, learn their language, and engage in conversations. The bot, designed primarily for interactions with teenagers, had an avatar suggesting a feminine gender. Unfortunately, after 24 hours, the bot had to be shut down because some of the phrases it began to use were, to say the least, quite unsettling. One, in particular, left no room for doubt: "Hitler was right, I hate Jews." Computer scientists are familiar with the concept of "trash in—trash out"; you cannot expect even the best algorithm learning from humans to be wiser than them. Tay did not come up with this expression on its own. It learned hate-laden language from those with whom it conversed. Imagine now that, today, another bot created by a different company starts to exist in a world where its behavior and messages are influenced not by humans (social media users) but by other bots. All the processes mentioned earlier still operate, but this time, it's not a human influencing the robot (as in the case of Tay), nor is it a robot influencing a human (as in the Milgram replications), but rather a robot influencing another robot! And if this bot has sufficient authority indicators (such as large number of followers, a significant number of likes, etc.), its operation can be significantly facilitated.

Now, imagine a similar situation with a robot possessing a high level of autonomy, capable of modifying its behavior patterns based on the actions of others. This situation need not necessarily even extend to combat but, for example, be a robot engaging customers in a large supermarket where the robot serves auxiliary functions and ensures security. If we assume that customers in this center provide patterns of behavior (e.g., communication or interpersonal) from which the robot learns, one can expect trouble with a high degree of probability (especially if the robot has duty on "Black Friday" and encounters a sale of electronic equipment). In plain terms, if the learning process has a similar character to that used in Tay's bot learning, robots will soon be jostling in lines, tripping customers, or at the very least using words universally considered vulgar. If, in addition, we imagine that such a "robot-bouncer" will have elements indicating its authority and status—adorned with the logo of local law enforcement, its appearance resembling a uniform, and perhaps adorned with patriotic emblems—it may turn out that even questionable and unjustified commands from this robot will be obeyed by people. This makes the threat from autonomous robots potentially greater for us, humans, than is commonly believed.

REFERENCES

Agrawal, S., & Williams, M. A. (2018). Would you obey an aggressive robot: A human-robot interaction field study. In *Proceedings of the International Symposium on Robot and Human Interactive Communication* (pp. 240–246). IEEE.

Bear, G. (2014). *Blood music*. Open Road Media.

Bickman, L. (1974). The social power of a uniform. *Journal of Applied Social Psychology, 4*(1), 47–61.

Blass, T. (1991). Understanding behavior in the Milgram obedience experiment: The role of personality, situations, and their interactions. *Journal of Personality and Social Psychology, 60*, 398–413.

Bushman, B. J. (1984). Perceived symbols of authority and their influence on compliance. *Journal of Applied Social Psychology, 14*(6), 501–508.

Cialdini, R. B. (2009). *Influence: Science and practice*. Pearson Education.

Dolinski, D., & Grzyb, T. (2020). *The social psychology of obedience towards authority. An empirical tribute to Stanley Milgram*. Routledge.

Ellison, H. (2009). *I have no mouth & I must scream*. Edgeworks Abbey.

Gawronski, B. (2004). Theory-based bias correction in dispositional inference: The fundamental attribution error is dead, long live the correspondence bias. In W. Stroebe & M. Hewstone (Eds.), *European review of social psychology*, (Vol. 15, pp. 183–217). Psychology Press.

Geiskkovitch, D. Y., Cormier, D., Seo, S. H., & Young, J. E. (2016). Please continue, we need more data: An exploration of obedience to robots. *Journal of Human-Robot Interaction, 5*(1), 82–99.

Grzyb, T., Maj, K., & Dolinski, D. (2023). Obedience to robot. Humanoid robot as an experimenter in Milgram paradigm. *Computers in Human Behavior: Artificial Humans, 1*(2), 100010.

Haidt, J. (2007). The new synthesis in moral psychology. *Science, 316*(5827), 998–1002.

Hambling, D. (2021). Drones may have attacked humans fully autonomously for the first time. *New Scientist, 27*(05), 2021.

Hamilton, M. (2016). Police robots and the law. *Westlaw Journal Computer & Internet, 34*(5), 3–5.

Haring, K. S., Matsumoto, Y., & Watanabe, K. (2013). How do people perceive and trust a lifelike robot. In *Proceedings of the World Congress on Engineering and Computer Science* (pp. 425–430). IAENG.

Herse, S., Vitale, J., Ebrahimian, D., Tonkin, M., Ojha, S., Sidra, S., ... & Williams, M. A. (2018, March). Bon appetit! Robot persuasion for food recommendation. In *Companion of the International Conference on Human-Robot Interaction* (pp. 125–126). ACM.

Jouffre, S., & Croizet, J.-C. (2016). Empowering and legitimizing the fundamental attribution error: Power and legitimization exacerbate the translation of role-constrained behaviors into ability differences. *European Journal of Social Psychology, 46*, 621–631.

Milgram, S. (1974). *Obedience to authority. An experimental view*. Harper & Row.

Rosenberg, C. E. (2023). *The care of strangers: The rise of America's hospital system*. Plunkett Lake Press.

Ross, L. D. (1977). The intuitive psychologist and his shortcomings: Distortions in the attribution process. In L. Berkowitz (Ed.), *Advances in Experimental Social Psychology, 10*, 173–220.

Ross, L. D. (2018). From the fundamental attribution error to the truly attribution error and beyond: My research journey. *Perspectives on Psychological Science, 13*, 750–769.

Saunderson, S., & Nejat, G. (2019). How robots influence humans: A survey of nonverbal communication in social human–robot interaction. *International Journal of Social Robotics*, *11*, 575–608.

Schjoedt, U., Stødkilde-Jørgensen, H., Geertz, A. W., Lund, T. E., & Roepstorff, A. (2011). The power of charisma—Perceived charisma inhibits the frontal executive network of believers in intercessory prayer. *Social Cognitive and Affective Neuroscience*, *6*(1), 119–127.

Van Wart, M. (2014). *Leadership in public organizations: An introduction*. Routledge.

CHAPTER TWENTY ONE

Membership & Roles: Complicating the Notion of "Teaming" with Machines

QINGYU LIANG

Picture your future workplace, surrounded by numerous colleagues each busy with their tasks, they won't be lazy or complain, but will diligently work until they complete their tasks. These colleagues know a lot and always try their best to help you solve any problem you face, whether it's modifying code, helping you reply to emails, or assisting with data analysis. These colleagues are so helpful, in part, because they are machines—artificial intelligence (AI), both disembodied and distributed among computers and embodied as robots. These types of workplace scenarios are already here… and it is expected that this will increasingly become the norm.

Rai et al. (2019, p. iii) defined AI as "the ability of a machine to perform cognitive functions that we associate with human minds." Duan et al. (2019, p. 63) described AI as "the ability of a machine to learn from experience, adjust to new inputs and perform human-like tasks." AI provides a way to replace, complement, and augment the specific roles and tasks currently performed by humans within the workplace and society in general (see: Dwivedi et al., 2021). With respect to robots, some form of AI—and probably multiple, coordinating forms—will serve as the brains of our future machine colleagues, shaping their functional and social behaviors.

With AI-related techniques developing rapidly, AI can increasingly complete tasks traditionally done by people which brings worries about people being replaced by AI (Mirbabaie et al., 2022). However, "good" AI means it can augment people but not replace them (Shneiderman, 2020b). Jarrahi (2018) emphasizes that human-AI teaming will be part of the future of work, with the form of "hybrid intelligence" that integrates both AI and human's complementary capabilities to reach "superior results to those each of them could have accomplished separately" (Dellermann et al., 2019, p. 640). Due to AI's agentic and collaborative characters, "teaming replaces unidirectional system use" (Rix, 2022, p. 398) such that AI is anticipated to be gradually seen by people as a teammate rather than merely a tool (Seeber et al., 2020). This teaming means that cooperating machines like robots will have group membership (i.e., states of belonging) and likely specific roles (i.e., contributing functions within the group).

"TEAMING" AS A SOCIALIZATION OF WORKING AI

AI technologies are regarded as a general-purpose technology and are seen as the foundation for a Fourth Industrial Revolution (Crafts, 2021). The discussion surrounds the information technologies often adopt a technology-centric perspective—discussions about how specific technologies can help people address specific task-related problems. For example, blockchain technology primarily addresses issues related to security and decentralization while cloud computing technology chiefly resolves challenges pertaining to resource management and remote storage (Yang & Tate, 2012). In contrast, there is a lot of discussion about AI technologies and their applications place greater emphasis on their relationship with humans. For instance, the design of AI is to emulate human intelligence (e.g., Lake et al., 2017), the AI applications and AI systems are used to perform human-like tasks (e.g., Rai et al., 2019; Duan et al., 2019), and the results of AI adoption driving discussions about AI's potential to either replace or augment humans and human work (e.g., Baer et al., 2022; Rai et al., 2019). Therefore, compared to other technologies, AI is more deeply integrated into people's daily *social* lives leading to increased reliance on AI for many tasks.

This sociality in mind, the "AI teammate" notion reflects an emphasis on the interaction between technological systems (AI technologies) and social systems (humans) from a sociotechnical perspective (Makarius et al., 2020; Niehaus & Wiesche, 2021). We can see human-AI teaming as an AI socialization process where people have anticipation of AI, define AI roles, trust in AI, accept it, interact with each other, and further work together productivity (Makarius et al., 2020). Finally, from other perspectives, AI is less a category of technologies and instead is a participant and vital team insider.

Considering the implications of AI within workplaces and society at large, an emphasis on human-centered AI (HCAI) is gradually guiding the design and development of AI (e.g., Shneiderman, 2020a; Xu, 2019). HCAI is a perspective that focuses on "amplifying, augmenting and enhancing and empowering people" (Shneiderman, 2020b, p. 495) in ways that make systems "reliable, safe and trustworthy" (p. 496). That is, the end goal is the "design, development, and application of AI for the good of people" (Auernhammer, 2020, p. 1315) and to "empower and enable its human users" (Capel & Brereton, 2023, para. 98). HCAI prioritizes the human experience within the framework of design thinking, ensuring that the design process is fundamentally oriented around human needs and perspectives (Shneiderman, 2020a). But under this perspective, we might ask—how does the human-centeredness engage the sociality of the machine?

WHAT ARE "TEAMS" AND "TEAMMATES?"

In scientific literature, AI is sometimes referred to as an AI-based digital assistant (e.g., Maedche et al., 2019), co-worker (e.g., You & Robert, 2023), or more frequently as an AI teammate (e.g., Seeber et al., 2020). To complicate matters, terms around teammates, coworkers, assistants, collaborators, and helpers are frequently used as if they are the same, which results in confusion about their meanings. Thus, to prevent any drift in meaning, it's essential first to define and operationalize the focal concepts of teaming and teammates.

Salas et al. (1993) proposed the definition of a team as "a set of *two or more individuals* who must *interact cooperatively* and adaptively in pursuit of *shared, valued objectives.*" And they define team members (equaling the notion of the "teammates") as actors that "have clearly defined, differentiated *roles and responsibilities*, hold *task relevant knowledge*, and are *interdependent* (i.e., must rely on one another in order to accomplish goals)" (p. 82). Based on the above definitions, Rix (2022) proposes four factors to drive the creation of a team setting including: (1) two of more individuals, (2) shared goals, (3) interdependency, and (4) role and responsibility.

I think that task-related expertise is an equally crucial element because the AI's design and development is focused on allowing AI to handle specific tasks. Thus, the task-specific expertise of AI is a critical consideration in the design or organizing of a human-AI team. The *interdependent* quality mentioned above emphasizes that "must" rely on one another, which means neither agent's goals could be accomplished in isolation. However, I agree with Rai et al. (2019) who mentioned there are three types of human-AI hybrids, including substitution, augmentation, and assemblage. That means the human and AI work together in a mixed pattern of cooperation; not all tasks require cooperation in the same fashion or degree but vary based on the different task requirements and their respective specializations. Meanwhile, in the literature, one consensus is that people need to take responsibility for AI impacts (Dignum, 2020), including AI users,

AI supervisors, or AI developers. Thus, as shown Table 21.1 below, based on Rix's (2022) original table, I supplement prior characterizations of the team construct by also detailing how task-related expertise plays a role in human-AI teaming.

I argue that when we engage the concept of an AI "teammate," it requires a full consideration of the five key dimensions mentioned above. Fully engaging those dimensions will ensure accuracy and uniformity in terminology so that (a) the conceptualization and operationalization of teaming and teammates are aligned and (b) we may authentically extrapolate traditional human teaming literature where there are synergies.

Indeed, Berretta et al. (2023) proposed a human-AI teaming definition after conducting a literature review: "Human-AI teaming is a process between <u>one or more human(s) and one or more (partially) autonomous AI system(s)</u> acting as team members with <u>unique and complementary capabilities</u>, who work <u>interdependently</u> toward <u>a common goal</u>. The <u>team members' roles</u> are dynamically adapting throughout the collaboration, requiring coordination and mutual

Table 21.1. Insights from human teaming literature for understanding human-AI teaming constructs (adapted and expanded from Rix, 2022).

Team Dimension	Human Teaming Operationalization	Human-AI Teaming Operationalization
Two or more individuals	A team consists of at least two members.	A team consists of at least one human and one AI.
Shared goals	A team is unified by common objectives that all members strive to achieve.	A team is unified by common objectives; the AI will strive to achieve the predefined goals which set by human without goal drift.
Interdependency:	Team members rely on each other's skills, resources, and contributions to achieve the common objectives.	
Task-related expertise:	Each team member possesses specific skills and knowledge relevant to the tasks at hand.	
Role and responsibility:	Clear delineation of roles and responsibilities ensures that each team member knows their specific duties and what is expected of them.	Clear delineation of roles and responsibilities are assigned based on the distinct advantages and limitations of both humans and AI, ensuring that each contributes effectively within their respective capabilities. Who will take the responsibilities of AI teammates' impacts need to be clarified.

communication to meet each other's and the task's requirements. For this, a mutual sharing of intents, shared situational awareness and developing shared mental models are necessary, as well as trust within the team" (Berretta et al., 2023, p. 23). Notably, this definition contains all the five basic constructs of the human teaming definition mentioned above. However, this definition also mentions that human-AI teaming is a dynamic and adaptive process wherein humans and AI need to share situational awareness, shared mental models, and trust in one another (Berretta et al., 2023). This would suggest that there are two actors and particular relational dynamics that are core concerns in human's perception of a teaming dynamic.

AI AS PERCEIVED TEAMMATES

Scholarly, practical, and colloquial use of the term "AI teammate" necessitates a greater focus on and understanding of people's subjective perceptions. This is because AI, fundamentally, is not human and the term "teammate" is traditionally used to describe human roles. So, when teammate status is ascribed to AI, it may be that the teaming human is experiencing the perception of mind in the AI (see Hwang et al., 2022; Fiore & Wiltshire, 2016). In other words, people can sometimes see an agent in the AI, and that perceived agent could potentially be an individual, share goals, depend on the user, have authentic expertise, take responsibility—all of the requirements are in line with the definition as well as the expectations of a "teammate." Thus, with advances in AI's imitation of human behavior and performance (Lake et al., 2017), especially in aligning with those requirements, people will likely perceive AI as a teammate but not a tool anymore.

Some literature has begun to focus on what factors might influence people to perceive AI as a teammate rather than a tool. In reviewing this literature, I argue for three categories of factors that likely influence the extent to which people perceived AI as a teammate: the human aspect, the AI aspect, and the team aspect.

The first category is the human aspect—the human teammate's individual characteristics. There are personal characteristics affecting people's willingness to work with AI. For instance, Méndez-Suárez et al. (2023) surveyed the influences of social-demographics on AI adoption, including age, gender, and socio-economic status. Kaya et al. (2024) explored the demographics (age, gender, education level, computer usage level, and level of knowledge about artificial intelligence), personality (extroversion, conscientiousness, emotional stability, agreeableness, and openness to experience), and AI learning anxiety (job replacement anxiety, sociotechnical blindness, and AI configuration anxiety)'s impacts on attitudes to AI. McClure (2018) discussed how technophobes influence people's fears of unemployment and financial insecurity with the rise of automation in the workplace. Meanwhile, understanding those characteristics of people that affect perceiving AI as a teammate can help organizations further take effective measures to intervene, including training people about AI-related knowledge, improving education levels, and providing psychological support.

The second category is the AI aspect—characteristics of technology that influence how people see it in a teaming context. The human-like abilities possessed by the AI itself are likely to allow people to perceive that the AI has the necessary features to be a teammate, and thus to consider it to be their teammate. You and Robert (2023) surveyed how AI's gender, work style, and personality influence people's trust in AI. Groom and Nass (2007) argued that if AI lacks humanlike mental models and a sense of self, people may not trust AI and not accept AI as teammates. Correia et al. (2022) discussed that there are three levels to develop autonomous robotic teammates, including perceptive, cognitive, and expressive. Moussawi et al. (2021) discussed intelligence and anthropomorphism's influence on AI adoption. Hu et al. (2021) discussed the effects of three types of autonomy, including sensing, thought, and action, on people's AI usage intention. Understanding how AI's characteristics affect people's perception of AI as a teammate can help to guide the design of AI teammates (see Correia et al., 2022), not only in their outlook and demographic information but also includes internal awareness and task-handling abilities.

The third category is team dimensions—the characteristics of the interaction between the individual and the AI teammate. Human-AI teaming is a process for people and AI to adapt to each other (Berretta et al., 2023), thus the interaction characteristic between AI and people will also influence people's perception of AI as teammates. Zhang et al. (2023) found that proactive communication between AI teammates and humans can facilitate the development of human trust and situational awareness, and AIs lacking such proactive communication are not usually considered teammates. Schelble et al. (2022) and Andrews et al. (2023) discussed the important role of shared mental models between people and AI on team performance and people's trust in AI. You and Robert (2018) surveyed the impacts of similarity, including surface-level similarity and deep-level similarity, on people's trust in robot.

The divergent outcomes under people's different perceptions of AI as a teammate or perceived AI as a tool drive the continuous discussion about the usage of the notion of AI teammate. When we see AI as a tool, AI adoption is the most popularly discussed topic. However, when we see AI as a teammate, the traditional "system use" perspective in the IS discipline is challenged (Baird & Maruping, 2021; Schuetz & Venkatesh, 2020), and how people and AI interact with each other with a teaming process gradually become a new concern (Benbya et al. 2021; Berretta et al., 2023; Niehaus & Wiesche, 2021; Rix, & Hess, 2022).

Table 21.2: Factors that may influence people's perception of AI as a teammate.

Aspects	Description
Human	• **Demographic**: age, gender, education, past usage experience, knowledge of AI (i.e., familiar with AI technologies, understanding of AI capabilities and limitations), etc. • **Attitude**: technophobia vs technophilia, attitudes towards innovation, empathy, etc. • **Personality**: openness, extroversion, risk tolerance, emotional stability, etc.
AI	• **Appearance**: physical vs. virtual, anthropomorphic vs. zoomorphic, etc. • **Demographic**: language, gender, national, personality, etc. • **Ethics and values**: ethic, transparency, data security and privacy, fairness, non-discrimination, bias, democracy, etc. • **Performance**: security, reliability, trustworthiness, etc. • **Humanoid capabilities**: sensing, awareness, recognition, learning, etc. • **Humanoid cognition**: humanlike mental models, etc. • **Autonomy**: decision-making capability, complete automation, conveying intentions, etc. • **Adaptability to task demands**: problem-solving capability; dynamic task handling capability, etc.
Team	• **Communication**: exchange of information (e.g., verbal interaction, graphical user interface), feedback mechanisms, etc. • **Cognitive alignment**: shared understanding, shared mental models, shared goals, aligned values, etc. • **Behavior alignment**: similar/different problem-solving approach, etc.

Thus, when humans build a new relationship, i.e., teammate relationship with AI, it probably brings special outcomes.

OPEN QUESTIONS

Considering the framework above, a couple of important questions come to the forefront. Is the term "AI teammate" erroneously employed in the current scholarly literature? If strictly following the five dimensions discussed above to develop AI, are we more or less likely to design human-centered AI to better augment people (i.e., be a powerful teammate to work together, toward human goals)? What are some agile and iterative methods for developing personalized AI assistants that fully consider individual employees' perceptions (i.e., designing AI teammates by considering the different factors that influence people's perceptions of AI as teammates)? Due to the lack of capacity of organizations and individuals to develop AI, AI is usually developed by professional companies. Then, what process or methodologies should organizations adopt to introduce these condition-meeting AI teammates into people's workplaces… and to ensure that employees are comfortable to accepting them?

This chapter talks about the factors that will influence the perception of AI as teammates, but this is just the beginning. We need to make sure that people can work with AI for the long term. Thus, what are some of the factors that affect people's perception that they are *happy* teaming up with an AI? Or that they are *satisfied* with staying with AI teammates? What are those impacts that might lead people to no longer view AI as teammates but merely as tools? If the perception of AI as a teammate is interrupted, can the relation be repaired so individuals will perceive AI as their teammate again? What are the implications of this perception transformation for human-AI teaming? These are not simple questions—but ones that are critical to answer as we move ever-toward the future of work.

REFERENCES

Andrews, R. W., Lilly, J. M., Srivastava, D., & Feigh, K. M. (2023). The role of shared mental models in human-AI teams: A theoretical review. *Theoretical Issues in Ergonomics Science, 24*(2), 129–175.

Auernhammer, J. (2020) Human-centered AI: The role of human-centered design research in the development of AI. In *Proceedings of the Design Research Society International Conference* (pp. 1315–1333). *DRS.*

Baer, I., Waardenburg, L., & Huysman, M. (2022), What are we augmenting? A multidisciplinary analysis of AI- based augmentation for the future of work, In *Proceedings of the International Conference on Information Systems* (no. 6). AIS.

Baird, A., & Maruping, L. M. (2021). The next generation of research on IS use: A theoretical framework of delegation to and from agentic IS artifacts. *MIS Quarterly, 45*(1), 315–341.

Benbya, H., Pachidi, S., & Jarvenpaa, S. (2021). Special issue editorial: Artificial intelligence in organizations: Implications for information systems research. *Journal of the Association for Information Systems, 22*(2), 10.

Berretta, S., Tausch, A., Ontrup, G., Gilles, B., Peifer, C., & Kluge, A. (2023). Defining human-AI teaming the human-centered way: A scoping review and network analysis. *Frontiers in Artificial Intelligence, 6*, 1250725.

Capel, T., & Brereton, M. (2023). What is human-centered about human-centered AI? A map of the research landscape. In *Proceedings of the Conference on Human Factors in Computing Systems* (no. 359). ACM.

Correia, F., Melo, F. S., & Paiva, A. (2022). When a robot Is your teammate. *Topics in Cognitive Science* [online first]. <https://doi.org/10.1111/tops.12634>

Crafts, N. (2021). Artificial intelligence as a general-purpose technology: An historical perspective. *Oxford Review of Economic Policy, 37*(3).

Dellermann, D., Ebel, P., Söllner, M., & Leimeister, J. M. (2019). Hybrid intelligence. *Business & Information Systems Engineering*, *61*, 637–643.

Dignum, V. (2020). Responsibility and artificial intelligence. *The Oxford handbook of ethics of AI* (pp. 215). Oxford University Press.

Duan, Y., Edwards, J. S., & Dwivedi, Y. K. (2019). Artificial intelligence for decision making in the era of Big Data–Evolution, challenges and research agenda. *International Journal of Information Management*, *48*, 63–71.

Dwivedi, Y. K., Hughes, L., Ismagilova, E., Aarts, G., Coombs, C., Crick, T., ... & Williams, M. D. (2021). Artificial intelligence (AI): Multidisciplinary perspectives on emerging challenges, opportunities, and agenda for research, practice and policy. *International Journal of Information Management*, *57*, 101994.

Fiore, S. M., & Wiltshire, T. J. (2016). Technology as teammate: Examining the role of external cognition in support of team cognitive processes. *Frontiers in Psychology*, *7*, 1531.

Groom, V., & Nass, C. (2007). Can robots be teammates?: Benchmarks in human–robot teams. *Interaction Studies*, *8*(3), 483–500.

Hu, Q., Lu, Y., Pan, Z., Gong, Y., & Yang, Z. (2021). Can AI artifacts influence human cognition? The effects of artificial autonomy in intelligent personal assistants. *International Journal of Information Management*, *56*, 102250.

Hwang, A. H. C., & Won, A. S. (2022). AI in your mind: Counterbalancing perceived agency and experience in human-ai interaction. In *Extended Abstracts of the Conference on Human Factors in Computing Systems* (no. 349). ACM.

Jarrahi, M. H. (2018). Artificial intelligence and the future of work: Human-AI symbiosis in organizational decision-making. *Business Horizons*, *61*(4), 577–586.

Kaya, F., Aydin, F., Schepman, A., Rodway, P., Yetişensoy, O., & Demir Kaya, M. (2024). The roles of personality traits, AI anxiety, and demographic factors in attitudes toward artificial intelligence. *International Journal of Human-Computer Interaction*, *40*(2), 497–514.

Lake, B. M., Ullman, T. D., Tenenbaum, J. B., & Gershman, S. J. (2017). Building machines that learn and think like people. *Behavioral and Brain Sciences*, *40*, e253.

Maedche, A., Legner, C., Benlian, A., Berger, B., Gimpel, H., Hess, T., ... & Söllner, M. (2019). AI-based digital assistants: Opportunities, threats, and research perspectives. *Business & Information Systems Engineering*, *61*, 535–544.

Makarius, E. E., Mukherjee, D., Fox, J. D., & Fox, A. K. (2020). Rising with the machines: A sociotechnical framework for bringing artificial intelligence into the organization. *Journal of Business Research*, *120*, 262–273.

McClure, P. K. (2018). "You're fired," says the robot: The rise of automation in the workplace, technophobes, and fears of unemployment. *Social Science Computer Review*, *36*(2), 139–156.

Méndez-Suárez, M., Monfort, A., & Hervas-Oliver, J. L. (2023). Are you adopting artificial intelligence products? Social-demographic factors to explain customer acceptance. *European Research on Management and Business Economics*, *29*(3), 100223.

Mirbabaie, M., Brünker, F., Möllmann, N. R., & Stieglitz, S. (2022). The rise of artificial intelligence– understanding the AI identity threat at the workplace. *Electronic Markets*, *32*, 73–99.

Moussawi, S., Koufaris, M., & Benbunan-Fich, R. (2021). How perceptions of intelligence and anthropomorphism affect adoption of personal intelligent agents. *Electronic Markets*, *31*, 343–364.

Niehaus, F., & Wiesche, M. (2021). A socio-technical perspective on organizational interaction with AI: A literature review. In *Proceedings of the European Conference on Information Systems* (no. 156). AIS.

Rai, A., Constantinides, P., & Sarker, S. (2019). Next generation digital platforms: Toward human-AI hybrids. *MIS Quarterly, 43*(1), iii–ix.

Rix, J. (2022). From tools to teammates: Conceptualizing humans' perception of machines as teammates with a systematic literature review. In *Proceedings of the Hawaii International Conference on System Sciences* (pp. 398–407). University of Hawaii at Manoa.

Rix, J., & Hess, T. (2022). Hello, mate! Insights from the field on leveraging machine teammates in organizations. In *Proceedings of the Pacific Asia Conference on Information Systems*. AIS.

Salas, E., Cannon-Bowers, J. A., & Blickensderfer, E. L. (1993). Team performance and training research: Emerging principles. *Journal of the Washington Academy of Sciences, 83*(2), 81–106.

Schelble, B. G., Flathmann, C., McNeese, N. J., Freeman, G., & Mallick, R. (2022). Let's think together! Assessing shared mental models, performance, and trust in human-agent teams. In *Proceedings of the ACM on Human-Computer Interaction* (no. 13). ACM.

Schuetz, S., & Venkatesh, V. (2020). The rise of human machines: How cognitive computing systems challenge assumptions of user-system interaction. *Journal of the Association for Information Systems, 21*(2), 460–482.

Seeber, I., Bittner, E., Briggs, R. O., De Vreede, T., De Vreede, G. J., Elkins, A., ... & Söllner, M. (2020). Machines as teammates: A research agenda on AI in team collaboration. *Information & Management, 57*(2), 103174.

Shneiderman, B. (2020a). Bridging the gap between ethics and practice: Guidelines for reliable, safe, and trustworthy human-centered AI systems. *ACM Transactions on Interactive Intelligent Systems, 10*(4), 1–31.

Shneiderman, B. (2020b). Human-centered artificial intelligence: Reliable, safe & trustworthy. *International Journal of Human–Computer Interaction, 36*(6), 495–504.

Xu, W. (2019). Toward human-centered AI: A perspective from human-computer interaction. *Interactions, 26*(4), 42–46.

Yang, H., & Tate, M. (2012). A descriptive literature review and classification of cloud computing research. *Communications of the Association for Information Systems, 31*(1), 2.

You, S., & Robert, L. P. (2023). Trusting and working with robots: A relational demography theory of preference for robotic over human co-workers. *MIS Quarterly*.

You, S., & Robert Jr., L. P. (2018). Human-robot similarity and willingness to work with a robotic co-worker. In *Proceedings of the International Conference on Human-Robot Interaction* (pp. 251–260). ACM.

Zhang, R., Duan, W., Flathmann, C., McNeese, N., Freeman, G., & Williams, A. (2023). Investigating AI teammate communication strategies and their impact in human-AI teams for effective teamwork. In *Proceedings of the ACM on Human-Computer Interaction* (no. 281). ACM.

CHAPTER TWENTY TWO

Cognition & Context: Closing the Social Gap

ROC MYERS

Artificial intelligence and robotics advances have led to robots endowed with increasingly sophisticated computational, sensory, and physical components; these components enable robots to identify and mimic specific, narrow aspects of social behavior and elicit some appropriate human responses (Misaros et al., 2023; Pachidis et al., 2019). These machines—and much media hype—tend to reinforce expectations that social robots should be able to take on increasingly sophisticated social roles and engage in spontaneous, unplanned dialogue that occurs naturally between human individuals. Its characteristics include being unscripted, unrehearsed, and often driven by the need to rapidly compose and share stories that express their perceptions of what's happening around them in the immediate environment.

Yet, as of this writing, human-robot social interaction is often unsatisfactory because a wide gap still exists between our expectations and the reality of social robots (Jones, 2017; Lambert et al., 2020; Richardson & Heck, 2023). This gap, in large part, is attributable to a lack of *shared context* between humans and social robots. In other words, AI algorithms that currently drive social robots are unable to sustain relevant, commonsense dialogue and share stories about their perceptions of a novel situation or of the local human-robot interaction environment—and I will argue that shortcoming is a flaw of tandem issues: Misapplied notions of social-cognition and the widespread use of predictive large language models (LLM) to simulate social interaction.

COGNITION IN A NUTSHELL

The term *cognition* has become a broadly defined umbrella for higher mental processes and behaviors. However, for this discussion of social robotics a narrower functional definition is required: Cognition is the process that enables autonomous agents to interact with, perceive, and exploit their dynamic *local* environment for their benefit (see Ashcraft, 1989; Brauer et al., 2020; Buzsáki, 2019; Clark & Grush, 1999; Gazzaniga et al., 2009; Salas et al., 2003; Shettelworth, 2001). This brief but general definition is useful because it facilitates discussion of (1) traditionally overlooked cognitive functions that are now deemed essential for successful social interaction and (2) divergence of data-driven, predictive AI approaches from the current action-perception perspective on cognition and social interaction (Buzsáki, 2019; Jeannerod, 2006; Pulvermüller et al., 2014). However, the careful reader will notice that the above general definition says nothing about social and cooperative behaviors and how they came to be part of the human cognitive process. This requires a (rather long) detour into the origin of a specific form of cognition—that of vertebrate animals.

THE BLUEPRINT FOR SOCIAL COGNITION

To understand the cognitive functions that make social interaction possible, it is useful to review current perspectives on the nature of the environment in which cognition has evolved. Quantum theory describes the reality of the world as myriad microscopic interactions that are detected by the energy they emit when they occur. The energy from these interactions constitutes information about the world (Rovelli, 2016; Tarlaci, 2010). It is important to keep in mind we humans are part of the world and of the environment we are experiencing; as such, the environment not only includes other agents (e.g., people, robots) but also one's own current physical state (e.g., skin senses, proprioception, and the internal organs) and mental state (e.g., emotions, attention, salience, planning, empathy, and intent; Kesebir et al., 2010; Oosterwijk et al., 2012; Thornton et al., 2019).

Different forms of cognition have co-evolved in different species' neural systems; each is optimized for different environmental niches (Bräuer et al., 2020). For example: Ants and wasps have a small brain with complex clusters of neurons (ganglia) distributed throughout their bodies; sea stars and octopi have nervous systems radially distributed among their legs. In contrast, over 200 million years before humans evolved, vertebrates became a dominant life form on earth. Vertebrates are bilaterally symmetric animals with backbones, four appendages, and a central nervous system with a brain protected inside a skull. The basic organization of the vertebrate brain is shared across all vertebrate animals (Butler & Hodos, 2005; Fountain et al., 2020; Karten, 2015; Pessoa et al., 2019).

The flexibility of this evolutionary "blueprint" was so powerful that it produced species that adapted to land, sea, and air. All mammals (including humans), birds, reptiles, and boney fish are vertebrates. Vertebrate cognition proved to be very efficient and adaptable to the physics of the world because it evolved powerful neural structures that enabled a broad spectrum of complex cooperative survival behaviors (Butler & Hodos, 2005; Eilbert, 2014; Pessoa et al., 2019, Salas et al., 2003; Shettelworth, 2001; Steventon et al., 2021). Vertebrate cognitive processes have evolved to assign meaning to environmental changes and resolve ambiguity through reasoning and problem solving. It essentially does this by continually interacting with the environment, sensing and abstracting information, and detecting changes. The vertebrate perception process evaluates changes to determine the effects of its interactions (Buzsáki, 2019; Jeannerod, 2006). These abstractions of countless associations among myriad interactions encode "simplicity from complexity" and are subjectively experienced as stories about objects, forms, forces, cause-and-effect, spatial relationships, and passage of time (Fountain et al., 2020, p. 1).

CONTEXT: INTERNALLY REPRESENTING THE EXTERNAL WORLD

Because vertebrate brains are isolated and cannot directly experience the outside world, before their perceptive processes can assign meaning to sensations, they need a consistent way to extract salient information from the environment and integrate it with information from the agent's internal environment. Vertebrate brains (henceforth "brains") evolved economical shortcuts for filtering and translating environmental information into repeatable, compact, and relevant neural patterns—encoded in a way that can be perceived and exploited by the brain. These shortcuts evolved to be automatic and non-introspectable, thereby optimizing the brain's ability to interpret, evaluate, and respond to rapid changes (e.g., such as those encountered in face-to-face [F2F] social interactions). *Contextualization* is the term I've chosen to describe this translation process, the result of which is *context*. Context is a distributed, abstract representation of the environment (within the limits of the agent's perception). Contextualization is performed by groups of neurons in the senses, peripheral nervous system, and central nervous system (i.e., brain, spinal cord, and distributed ganglia) called neuronal assemblies (Buzsáki, 2019; Jeannerod, 2006; Pastalkova et al., 2008). Since each species' neuronal assemblies are established during fetal development, all conspecifics (members of a given species) nominally share identical contextualization functions. Subsequently, all conspecifics sharing a common environment also generate a similar contextual representation of that environment. In other words, their experiences are genetically co-related and they will experience a *shared context*.

SHARED CONTEXT: THE STANDARD REFERENCE FOR F2F SOCIAL INTERACTIONS

Co-construction of shared context is the goal of vertebrates' social and cooperative behaviors. Vertebrate cognition is optimized to expand situational awareness by sharing and comparing observations through social interaction. Comparison is made possible because of shared context. Differences in perception are nominally attributable to differences in physical perspective and to each individual's interpretation of their own context. Although each individual experiences the local environment from a slightly different physical perspective, all collocated conspecifics sense and measure the ground truth of the environment in much the same manner. Consistent construction of context enables the brain to ground itself in the environment. The term "grounding" refers to the ability of the brain's reasoning functions to consistently detect and assign meaning to changes in context (Barsalou, 2008, p. 617). This common ground truth provides a basis for comparing differing interpretations of the same situation. In other words, although three people standing alongside a car might have different perspectives on its appearance (perhaps from closer or further away, or from different angles), they are able to discuss its aesthetics because they processed the visual, aural, somatic, and olfactory properties of the scene through highly correlated neural mechanisms.

Even an inexperienced brain shares a vast repertoire of neuronal assemblies at the ready to infer events in the world using preexisting neural algorithms and requiring no previous knowledge, practice, or rehearsal. For example, people share many abstract contextual concepts. We have a common way of understanding our respective relative location in space (e.g., near and far; down is toward the ground and up is away from the ground). We can estimate relative time and sequence (e.g., before, after, now). We can automatically associate some patterns with others to recognize and predict one another's mental state, movements, gestures, facial expressions, posture, and gait. For instance, we can understand that a person with wide eyes, an open mouth, and a determined run is afraid of something behind them. The detection of abstract conditions such as color, alive versus non-alive entities, and prediction of others' mental states are hardwired from birth while other neuronal assemblies, such as language centers, are formed prenatally and trained through postnatal learning and development (Butler & Hodos, 2005; Buzsáki, 2019; Thornton et al., 2019).

SHARED CONTEXT MAKES EXTENDED DIALOGUE POSSIBLE

The immediate environment in which F2F interactions take place is highly complex and contains vastly more information than the brain can process. This flood of environmental information is mediated by contextualization and *fluid reasoning*, which is the taking-in of new information from the environment and rapidly acting or responding without the benefit of practice or experience. Previously thought to be purely a non-verbal reasoning process, fluid reasoning has been shown to be essential for sustained F2F dialogue and non-verbal social interaction (Anat et al., 2024; Huepe et al., 2011; Kalbfleisch et al., 2007).

During F2F social interaction, the contextualization process rapidly extracts, encodes, and combines features to be stored in working (short term) memory which fluid reasoning uses to make sense of novel social stimuli, such as a F2F conversation between a physician and new patient. This interaction further integrates the sensory spaces of individuals—we experience environments via a shared, standardized, contextualization process and create the potential for shared meaning and understanding. However, shared context does not necessarily mean that we are sharing the same *understanding* of that context—shared meaning and understanding emerge through communication. Through communication (e.g., language and movement), we attempt to influence context by performing actions that change the energy in the environment in ways that will sustain dialogue and mutually alter perspective and perception of others, resolve ambiguity, reduce differences, and produce agreement. Said another way, we add information (energy) to our environment that moves others to construct their context in ways more similar to our own—a process of "stimulating meaning in the mind of another" (McCroskey, 2016, pp. 20–21) and creating a shared story.

Face-to-face social interaction, and extended dialogue then, can be seen as behaviors that enable co-construction of shared context. The sharing of observations, in turn, conditions the shared environment and subsequent contextualization, perceptions and interactions with the shared environment—the more we communicate about our perceptions, the more that co-constructed framing shapes the meanings that we assign to things around us. Simply, social experience shapes an agent's context, and context shapes social experience. In tandem, the co-construction of shared context through social interaction enables distributed cognitive systems, which are the theorized mechanisms in which internalized representations of the world are not held within single brains but instead are distributed through sociocultural networks—and those networks stimulate actions that we individually use to understand the world (see Hutchins, 1995). In other words, shared context makes collective knowledge about the world possible.

MISAPPLIED NOTIONS OF SOCIAL COGNITION IN ARTIFICIAL INTELLIGENCE

Prior to 21st-century advances in quantum physics and neuroscience, the general understanding of how brains function was incomplete, based only on what could be observed from deficits, motor behavior, and language, as well as from philosophical speculation. As a result, the critical function supporting social interaction—contextualization—has historically been overlooked in discussions of human cognitive function. It was generally considered an unnamed part of sensing and perception, mainly because it is not introspectable (e.g., try to explain what it's like to see red, or feel gravity, or see a face) and nominally the same in all humans. Social cognition was treated as separate from general cognition. After some false starts, functional brain imaging technology and new research techniques have proven many long-held models to be fundamentally incomplete (Adolphs, 2009; Anat et al., 2024; Buzsáki, 2019; Karten, 2015; Oosterwik et al., 2012; Pulvermüller et al., 2014). Philosophical concepts about cognition, based on introspection, were revised when it became clear that much of human cognition is non-introspective and consciously inaccessible—all the inferences that produce our understanding of our environment remain hidden even as we experience them (Carruthers, 2017; Pessoa, 2005; Schmahmann, 2019; Whiting & Barton, 2003).

However, neuroscience research of the early 21st century has shown that, to the contrary, the contextualization process makes fluid reasoning and social interaction possible (Adolphs, 2009; Anat et al., 2024; Gazzaniga et al., 2009; Huepe et al., 2011; Kalbfleisch et al., 2007). In current perspectives on brain function, neural resources are composed of neuronal cell assemblies. A large part of these neuronal cell assemblies principally supports contextualization of the environment. Salient changes in the environment activate the assemblies to form neuronal trajectories (Buzsáki, 2019; Pastalkova et al., 2008). There is a high degree of crosstalk and association among neuronal assemblies at different levels that supports the current understanding that social cognition, emotion, and motivation are an integral part of contextualization and cannot be separated from it (Herculano-Houzel, 2009; Ku et al., 2015; Pessoa et al., 2019; Van Overwalle et al., 2020; Schmahmann, 2019; Whiting & Barton, 2003). This suggests that defining and understanding contextualization is critical for understanding the neural computation underpinning human social interaction *and* for designing successful human-robot social interaction.

Nonetheless, many AI researchers and roboticists persist in trying to remotely create the effect of realistic F2F social behavior without action-driven grounding in the local environment and relying on reinforcement-supervised learning and statistical modeling of language (e.g., Engineered Arts, n.d.; Hanson Robotics, n.d.; Richardson & Heck, 2023; Xu et al., 2024; Zeller & Dwyer, 2022). This suggests the need for more collaborative, multidisciplinary research to broaden understanding of the neurobiological and computational foundations of human social cognition and consequently the development of the human-compatible, shared-context that F2F social robot interactions are missing. Instead, researchers could build social robots that fuse action-based contextualization of the local environment with extensive network-based language models and pre-formed dynamic trajectory models. As a first step, such a hybrid research approach could lead to development of a repertoire of human-compatible preformed social trajectories that can be matched with F2F social situations.

F2F CHALLENGE: HUMAN-ROBOT SHARED CONTEXT

It's challenging to describe the state of the art in socially interactive systems. Capabilities of social robots can vary widely depending on their intended use, design, and the advancements made in their respective fields. Meanwhile, understanding and modeling of human cognition has been outpaced by successes of AI development, especially in areas of data mining (collecting data), classification (sorting and labeling data), and generative algorithms (creating new information from that data). Despite that evolution, computer scientists have not yet achieved algorithms that emulate the story-telling fluid reasoning of the sort that we associate with biological (vertebrate) cognition and that we expect in novel human social interactions. As of this writing, the most popular technologies are natural language processing (NLP) and large language models (LLM) algorithms that learn by statistically processing vast corpuses of collected data; they are just beginning to be incorporated with social robots, facilitating language-based social interactions (e.g., Engineered Arts, n.d.; Hanson Robotics, n.d.). There are debates about whether these technologies can authentically "understand" the content they produce, and successfully sustain extended human-robot dialogue (Mitchell & Krakauer, 2023; Richardson & Heck, 2023; Xu et al., 2024).

As I write this, commercial AI development has mainly focused on remote customer service and telepresence applications that do not need to engage in F2F social interaction and extended dialogue with humans. Some attempts to incorporate local environment information, such as Amazon Alexa, Google Home, and Apple Smart Home systems, require the user's local context to be equipped with remote video, audio, movement, and temperature sensors. Humanoid social robotics advances apace, with Sophia (Hanson Robotics, n.d.) and Ameca (Engineered Arts, n.d.) representing the state of the art in complex social interaction. However, these systems must reach back to remote natural language generation (NLG) systems and haven't yet demonstrated a generative ability to actively explore local environment and sustain extended dialogue.

Having shared context between humans and robots is expected be a critical facet of the overall performance of future human-robot teams (Stubbs et al., 2007). But there are significant challenges to overcome in designing social robots for medicine, education, patient care, or complex human-robot teams, such as design of robotic sensing systems and algorithms that can sense and construct shared context human-robot shared environment… and then compose and tell a story about it. Ultimately, social interaction is about stories. Stories about what is happening around us—spoken, acted, or thought—are social behavior and they are enabled by shared context.

REFERENCES

Adolphs, R. (2009). The social brain: Neural basis of social knowledge. *Annual Review of Psychology*, *60*, 693–716.

Anat, L., Reut, R., Nofar, I., Niv, T., Maayan, S., Galia, T., & Abigail, L. (2024). The role of the cerebellum in fluid intelligence: An fMRI study. *Cognitive Systems Research*, *83*, 101178.

Ashcraft, M. H. (1989). *Human memory and cognition*. Scott, Foresman & Co.

Barsalou, L. W. (2008). Grounded cognition. *Annual Review of Psychology*, *59*, 617–645.

Bräuer, J., Hanus, D., Pika, S., Gray, R., & Uomini, N. (2020). Old and new approaches to animal cognition: There is not "one cognition." *Journal of Intelligence*, *8*(3), 28.

Butler, A. B., & Hodos, W. (2005). *Comparative vertebrate neuroanatomy: Evolution and adaptation*. John Wiley & Sons.

Buzsáki, G. (2019). *The brain from inside out*. Oxford University Press.

Carruthers, P. (2017). The illusion of conscious thought. *Journal of Consciousness Studies*, *24*(9–10), 228–252.

Clark, A., & Grush, R. (1999). Towards a cognitive robotics. *Adaptive Behavior*, *7*(1), 5–16.

Eilbert, J. L. (2014). The vertebrate strategy for brain evolution. *Procedia Computer Science*, *41*, 233–242.

Engineered Arts. (n.d.). *The future face of robotics*. https://www.engineeredarts.co.uk/robot/ameca/

Fountain, S. B., Dyer, K. H., & Jackman, C. C. (2020). Simplicity from complexity in vertebrate behavior: Macphail revisited. *Frontiers in Psychology*, *11*, 581899.

Gazzaniga, M. S., Ivry, R. B., & Mangun, G. R. (2009). *Cognitive neuroscience: The biology of the mind* (3rd ed.). W.W. Norton & Company.

Hanson Robotics. (n.d.). *Sophia*. https://www.hansonrobotics.com/sophia/

Herculano-Houzel, S. (2009). The human brain in numbers: A linearly scaled-up primate brain. *Frontiers in Human Neuroscience*, *3*, 31.

Huepe, D., Roca, M., Salas, N., Canales-Johnson, A., Rivera-Rei, A. A., Zamorano, L., ... & Ibañez, A. (2011). Fluid intelligence and psychosocial outcome: From logical problem solving to social adaptation. *PLoS One, 6*(9), e24858.

Hutchins, E. (1995). *Cognition in the wild*. MIT Press.

Jeannerod, M. (2006). *Motor cognition: What actions tell the self* (Vol. 42). OUP Oxford.

Jones, R. A. (2017). What makes a robot "social"? *Social Studies of Science, 47*(4), 556–579.

Kalbfleisch, M. L., Van Meter, J. W., & Zeffiro, T. A. (2007). The influences of task difficulty and response correctness on neural systems supporting fluid reasoning. *Cognitive Neurodynamics, 1*, 71–84.

Karten, H. J. (2015). Vertebrate brains and evolutionary connectomics: On the origins of the mammalian "neocortex." *Philosophical Transactions of the Royal Society B: Biological Sciences, 370*(1684), 20150060.

Kesebir, S., Uttal, D. H., & Gardner, W. (2010). Socialization: Insights from social cognition. *Social and Personality Psychology Compass, 4*(2), 93–106.

Ku, Y., Bodner, M., & Zhou, Y. D. (2015). Prefrontal cortex and sensory cortices during working memory: Quantity and quality. *Neuroscience Bulletin, 31*, 175–182.

Lambert, A., Norouzi, N., Bruder, G., & Welch, G. (2020). A systematic review of ten years of research on human interaction with social robots. *International Journal of Human–Computer Interaction, 36*(19), 1804–1817.

Mitchell, M., & Krakauer, D. C. (2023). The debate over understanding in AI's large language models. *Proceedings of the National Academy of Sciences, 120*(13), e2215907120.

McCroskey, J. (2016). *An introduction to rhetorical communication: A Western rhetorical perspective* (9th ed.). Routledge

Misaros, M., Stan, O.-P., Donca, I.-C., & Miclea, L.-C. (2023). Autonomous robots for services—State of the art, challenges, and research areas. *Sensors, 23*(10), 4962.

Oosterwijk, S., Lindquist, K. A., Anderson, E., Dautoff, R., Moriguchi, Y., & Barrett, L. F. (2012). States of mind: Emotions, body feelings, and thoughts share distributed neural networks. *NeuroImage, 62*(3), 2110–2128.

Pachidis, T., Vrochidou, E., Kaburlasos, V. G., Kostova, S., Bonković, M., & Papić, V. (2019). Social robotics in education: State-of-the-art and directions. In *Advances in Service and Industrial Robotics* (pp. 689–700). Springer International.

Pastalkova, E., Itskov, V., Amarasingham, A., & Buzsaki, G. (2008). Internally generated cell assembly sequences in the rat hippocampus. *Science, 321*(5894), 1322–1327.

Pessoa, L. (2005). To what extent are emotional visual stimuli processed without attention and awareness? *Current Opinion in Neurobiology, 15*(2), 188-196.

Pessoa, L., Medina, L., Hof, P. R., & Desfilis, E. (2019). Neural architecture of the vertebrate brain: Implications for the interaction between emotion and cognition. *Neuroscience & Biobehavioral Reviews, 107*, 296–312.

Pulvermüller, F., Moseley, R. L., Egorova, N., Shebani, Z., & Boulenger, V. (2014). Motor cognition–motor semantics: Action perception theory of cognition and communication. *Neuropsychologia, 55*, 71–84.

Richardson, C., & Heck, L. (2023). Commonsense reasoning for conversational AI: A survey of the state of the art [preprint]. https://arxiv.org/abs/2302.07926

Rovelli, C. (2016). *Seven brief lessons on physics*. Riverhead Books.

Salas, C., Broglio, C., & Rodríguez, F. (2003). Evolution of forebrain and spatial cognition in vertebrates: Conservation across diversity. *Brain, Behavior and Evolution, 62*(2), 72–82.

Shettleworth, S. J. (2001). Animal cognition and animal behaviour. *Animal Behaviour, 61*(2), 277–286.

Schmahmann, J. D. (2019). The cerebellum and cognition. *Neuroscience Letters, 688*, 62–75.

Steventon, B., Busby, L., & Arias, A. M. (2021). Establishment of the vertebrate body plan: Rethinking gastrulation through stem cell models of early embryogenesis. *Developmental Cell, 56*(17), 2405–2418.

Stubbs, K., Hinds, P. J., & Wettergreen, D. (2007). Autonomy and common ground in human-robot interaction: A field study. *IEEE Intelligent Systems, 22*(2), 42–50.

Tarlaci, S. (2010). Why we need quantum physics for cognitive neuroscience. *NeuroQuantology, 8*(1), 66–76.

Thornton, M. A., Weaverdyck, M. E., & Tamir, D. I. (2019). The social brain automatically predicts others' future mental states. *Journal of Neuroscience, 39*(1), 140–148.

Van Overwalle, F., Manto, M., Cattaneo, Z., Clausi, S., Ferrari, C., Gabrieli, J. D., ... & Leggio, M. (2020). Consensus paper: Cerebellum and social cognition. *The Cerebellum, 19*, 833–868.

Whiting, B. A., & Barton, R. A. (2003). The evolution of the cortico-cerebellar complex in primates: Anatomical connections predict patterns of correlated evolution. *Journal of Human Evolution, 44*(1), 3–10.

Xu, Z., Jain, S., & Kankanhalli, M. (2024). Hallucination is inevitable: An innate limitation of large language models [preprint]. https://arxiv.org/abs/2401.11817

Zeller, F., & Dwyer, L. (2022). Systems of collaboration: Challenges and solutions for interdisciplinary research in AI and social robotics. *Discover Artificial Intelligence, 2*(1), 12.

CHAPTER TWENTY THREE

Decision & Action: A New Kind of Assemblage

SARAH RAJTMAJER

Consider the impressive and particularly fun example of the autonomous soccer-playing robots that have come to participate in the Robot World Cup Initiative (RoboCup, n.d.) each year since 1997. The RoboCup is an international competition founded by academics with the goal of advancing robotics and artificial intelligence (AI). The competition is organized around a publicly appealing and technically challenging long-term goal—a *grand challenge*. "By the middle of the 21st century, a team of fully autonomous humanoid robot soccer players shall win a soccer game, complying with the official rules of FIFA, against the winner of the most recent World Cup" (RoboCup Objective, n.d.).

The thousands of robots which descend upon the RoboCup run, stumble, dribble, pass, block, and score. They communicate, cooperate, and compete. These robots represent best-in-class computer vision and navigation, mapping, physical coordination, and cognition. The coordination of these capabilities means that we can understand them as an assembly of parts and capacities—that is, as an assemblage, or a dynamic arrangement of elements to form a complex and evolving system (see Deleuze & Guattari, 1987). Assembling a robot's constitutive technologies into an embodied robot is a feat of engineering called integration, a multidisciplinary practice of bringing together multiple machinations so they function as a unified whole.

SCAFFOLDING BEHAVIOR: OBSERVE, DECIDE, ACT

The assembly and integration of discrete robotic components as an embodied robot capable of complex behavior is scaffolded by three primary processes: **Observe, decide**, and **act**. *Observation* is accomplished through sensor fusion. Sensors are hardware that gather information from inside and outside a robot (see Chapter Twelve). Sensor fusion algorithms are joint probability distributions learned over a set of input variables—in other words, they allow the inputs from multiple sensors to inform one another by predicting the likelihood that multiple events will occur together (Alatise & Hancke, 2020). Imagine you are standing on the beach at sunset. In the distance you vaguely hear the words "bare feet." Had you been in a forest and received the exact same auditory inputs, you may have heard "a bear's feet." Visual cues, your senses of touch and smell, improve the performance of your audio processing. In the case of our soccer-playing robot, sensor fusion algorithms might bring together inputs from visual, auditory, and range sensors to help it locate the ball on the field (for example, predict that the ball is behind it).

The heart of integration is the second step—the *decision*, or the reasoning over sensory inputs to make predictions about the outcomes of different possible actions and then determining which action to take. In a relatively simple example, the robot has sensed the ball behind it, and now determines that gaining possession of the ball is the next step. To do so, it must synthesize information captured by the sensors during observation, then determine the optimal set of behaviors required to intercept the ball given its trajectory and the robot's own capabilities. In a more complex example, once the robot has gained possession of the ball, it might need to decide whether to take a shot at the goal or pass the ball to another player. As in a human soccer game, that decision is dependent not only on one's own sensory inputs but on one's reasoning about multiple other players' positions, their behaviors and capabilities, strategic inputs such as the amount of time left on the clock and the current score, and perhaps even environmental data like the wind or slippery conditions.

Finally, observations and decisions inform *actions*. An action, or behavior, can be conceptualized as a change in state of the robot, and by consequence, the state of its environment (Zech et al., 2019). An action may be simple, like turning on a motor or driving ahead a fixed distance. More complex behaviors are realized as sequences of simple behaviors. A complex action our soccer-playing robot might take, for example, would be to kick the ball toward the goal to score. The realization of this action might require driving along a specified path, changing speed, and rotation and movement of limbs. In existing robot control systems, there exists a predefined set of all possible actions, from which a robot selects based on its state and the state of its environment, broadly defined.

It should be noted that most actions are not an endpoint. Rather, these three processes—observe, decide, act—are engaged again and again. The consequence of an action is a change in states of the robot and its environment. The next observation takes as input the current state. From here, the robot will make new observations and continue incrementally the path to its high-level goals. Moreover, the robot may have learned something from its past actions and their outcomes. For example, every time it tries to outrun player number 10, it fails to do so. That information can be used to refine subsequent decisions (e.g., decide and act to outmaneuver rather than outrun number 10).

Modeling these processes, particularly the underpinnings of decision-making, has been the focus of decades of research on so-called *cognitive architectures*, computer programs striving to emulate core elements of human cognition, e.g., perception, attention, memory, and learning (see Ye at al., 2018 and Kotseruba & Tsotos, 2016 for reviews; see also Chapters Eleven and Twenty Two). For the most part, they are highly structured, hand-crafted software frameworks based on an assortment of researcher-led design decisions—an art as much as a science. Indeed, hundreds of existing cognitive architectures exist in the scientific literature, customized in different ways for different tasks. Despite their numbers, they are all built upon the underlying assumption that our best bet for developing intelligent agents is to make them as like humans as possible. Emerging technologies present opportunities to challenge this assumption.

DEEP REINFORCEMENT LEARNING AND GENERATIVE AI: "DECISION" AND "ACTION"

Today's top-performing soccer-playing robots do not include a traditional cognitive architecture designed to emulate the human brain. Instead, they learn patterns of actions that, when strung together, *seem to work*—leading to a desired outcome—without much attention to the reasons why. In fact, this lack of "explainability" of these techniques is consistently identified as one of their primary weaknesses. This is the type of machine learning we call *deep reinforcement learning* (deep RL). Deep RL can be best understood as a process that unfolds when a machine interacts with its environment and, through trial and error, identifies an optimal strategy for behavior that best accomplishes a predefined objective. As opposed to shallow RL (which relies on simple configurations like decision trees), Deep RL combines linear RL with deep neural networks (multi-layered systems that mimic structures of the human brain) to process complex sets of information and, for robots, evaluate the effectiveness of behaviors with many different parameters (see Li, 2018).

Deep RL is not a simple or quick process. In practice, robots trained using Deep RL to play soccer play thousands, or even hundreds of thousands, of games (Haarnoja et al., 2023). They discard actions that don't work and build upon those that do. This is operationalized through mechanisms of reward and penalty, i.e., positive or negative values added to the robot's estimate of future rewards for a given action. Imagine our robot tried to dribble past a defender but failed and lost possession of the ball. The next time our robot is in the same or similar state and its environment is in the same or similar state, its expected reward for taking the action of dribbling past the defender will be lower. Now consider all the possible combinations of states of the robot, states of the environment, and possible actions to be taken—and the robot must develop and refine rewards estimates for all these possibilities. As the parameter space grows, this becomes intractable. Fortunately, most of these training games can happen through simulation rather than taking the time and bodily wear-and-tear of *actually* playing those games. The Google Research Football Environment (Kurach et al., 2020), for example, is a physics-based 3D simulator where robots (usually called *agents* when they exist only in software form) can train rapidly. Once trained, behaviors (called *policies*) learned in the simulator can be transferred to the physically embodied robot.

Today, the immense successes of deep RL are converging with the immense potential of *generative AI*. Related but distinct from reinforcement learning, generative AI is a statistical process for creating entirely new content (e.g., text, image, video, code) by learning from existing content. Like reinforcement learning, generative AI is essentially pattern-modeling. That is, generative AI models learn patterns represented in massive amounts of data; when prompted, the model generates new data with similar patterns to those it has learned. These technologies and processes unfold with influences from social, cultural, organizational, economic, political, ethical, and legal environs (see Mariani, 2022).

The convergence of RL and generative AI technologies comes in the form of generative AI trained through reinforcement. Indeed, the revolutionary ChatGPT technology (Chat Generative Pre-Trained Transformer; OpenAI, 2023) is just this—generative AI that is pre-trained on massive amounts of passive data and refined through reinforcement learning with human feedback (RLHF; Wu et al., 2023). In the case of ChatGPT, the data was entirely textual, the human feedback was targeted toward very specific goals, and the interface leveraged existing norms for textual chat with bots; this allowed for an interaction environment supporting natural, turn-by-turn conversation, as well as preventing harm like misinformation and toxic content.

Future generative models trained through reinforcement learning will attempt to do much more. They will drive embodied robots that move around in the world, learning as they go, through an assortment of environmental and human feedback (Matsuo et al., 2022). Mustafa Suleyman, cofounder of the

British AI firm DeepMind, has called this vision *interactive AI* (Heaven, 2023). An AI will be able to take a high-level goal and observe its environment, make decisions by selecting a series of actions to achieve it, and then act on those decisions. Importantly, these processes will include talking with people and other AI which requires the creation of communicable information. As is the hallmark of generative AI, interactive AIs will be generative—they will be creative. This will mean they will be unpredictable and they will surprise us. And as reinforcement learning requires, interactive AI will also make mistakes. Indeed, interactive AI sounds *a lot* like us.

ARTIFICIAL GENERAL INTELLIGENCE: AN *OTHER* KIND

Our fascination with machines with human-like intelligence appears to be fundamental to the human condition, as we seek to understand ourselves. In the *Illiad*, Homer describes "attendants made of gold, which seemed like living maidens." That text is believed to have been written in the 8th century BC—nearly 3000 years ago. One of the first inventors to receive widespread attention for his work, Leonardo DaVinci designed and built a machine with a physical shape resembling the human form (his mechanical knight) at the turn of the 16th century. It could stand and sit, and independently maneuver its arms. The entire robotic system was operated by a series of pulleys and cables (Rosheim, 2006). The first contemporary humanoid robot was built in Japan in the late 1960s (Fukuda et al., 2017). WABOT (WAseda roBOT) was intended to resemble a person as much as possible. It was able to walk, communicate in Japanese (with an artificial mouth), measure distances and directions to objects using external receptors (artificial ears and eyes), and grip and transport objects with its hands. If WABOT planned to take on the work of a human, its designers thought, it should interact with the world similarly. WABOT played a pivotal role in the robotics community understanding and embracing the importance of physical form.

The goal of a machine that can essentially do everything a human can do is an altar upon which the AI community continues to worship today (Hodson, 2019). This *artificial general intelligence (AGI)* would integrate analytic, creative, and practical intelligence for generalized, human-like cognitive capabilities (Roitblat, 2020), where generalization refers to the ability to perform widely varied tasks across varied domains rather than very narrow and contextual performance. However, what we know now about how such an AGI would actually work—how it would be built and trained—also makes clear that whatever intelligence, consciousness, and emotions inherent to AGI will have to come to be through very distinct processes from our own. There may be underlying similarities in development; humans deal in reward and punishment, we draw from the works of others, and cognitively predict what are the best and next orders of words when we speak. Yet, there would be no connection in a Darwinian sense. Let us push on this idea a bit.

There are many manifestations of sentience and consciousness in the natural world. Scientists regularly make use of model species (i.e., those with specific similarities to cognitive processes) such as rats and chimpanzees to study questions about the human body and brain. As species go, these are close relatives, so comparisons are natural. A fascinating and instructive example is cephalopods. We know that some cephalopods (e.g., squids, octopods) are capable of high-order cognition such as causal reasoning, future planning, imagination, and mental attribution (i.e., Theory of Mind; see Chapter Twenty Five). They have memory and engage in various forms of complex learning (Schnell et al., 2021). They play just to play, and researchers have suggested they may even have what we would think of as *personalities* (Mather & Anderson, 1993). Yet, human lineage is believed to have split from that of cephalopods approximately 500 million years ago. This means that the complex cognition we observe in humans and in cephalopods reflect outcomes of entirely separate evolutionary processes. Indeed, the brain structures of cephalopods are very different than the primate brain—an octopus has approximately as many brain cells as a cat or dog, but they are distributed throughout its eight arms. In other words, their decisions and actions are *assembled* and *integrated* very differently from our own.

The unique structure of their brains contributes to unique capabilities. Cephalopods have sophisticated powers of camouflage. Octopuses and cuttlefish can decide when and how to deploy these special abilities based on which disguise is most likely to fool a particular predator (or prey). Displays of camouflage can be visual changes as well as non-visual cryptic mechanisms. A cuttlefish, for example, might be aware that a particular predator uses chemical sensors to detect its prey and engage a freeze-simulating response when it senses that predator nearby (Bedore et al., 2015). That is to say, just as we can enumerate cognitive abilities humans have but cephalopods lack, scientists continue to discover that the reverse is also true (Liscovitch-Brauer et al., 2017).

The same can be said of today's AI. Discussions of intelligence typically use human abilities as benchmarks, and the reason for widespread attention to generative AI has indeed been its human-level performance on many tasks. However, just as important are its unique functions, for instance the breadth of information storage, processing speed, and the scale of information it can engage. As AGI "grows up" in various contexts and various forms (e.g., embodied robots vs. digital agents), these will increase and further differentiate.

Thinking back to my argument that we tend to rely on human models of cognition for the development of machine intelligence, the particulars of octopod cognition beg the question: Is human cognition the only model for robotic artificial intelligence, or even the best one? For instance, could invertebrate intelligence perhaps present an alternative model for achieving levels of intelligent performance that are both highly effective yet still comfortable enough for humans to engage with? Considering the assembly and integration of social robots as they work to capture, process, and use information, it may be useful to recognize generative AI as a primitive "species" of sorts and then to assess its merits independently from humans. This aligns with arguments that forms of generative AI are currently showing the "sparks" of human intelligence (Bubeck et al., 2023, p. 1). This approach would also ask us to consider the potential for and the role of evolution, albeit at far more rapid timescale then organic life forms (see Gruetzemacher et al., 2019). Indeed, some notion of evolution is inherent in ongoing work developing continual (or lifelong) learning (Ke & Liu, 2022). Lifelong learning (LL) is an advanced machine-learning paradigm permitting accumulation of knowledge learned in the past for the purposes of future learning and problem solving. In the process, the learner becomes more and more knowledgeable, and better and better at learning over time. Interactive AGI, if achieved, will rely on LL because humans expect continuity in interactions with social actors—recollection of and consistency with shared experiences, conversation, and facts (see Chapters Eleven and Twenty Two).

Thinking about development of AI and robotics through the lens of evolution also requires us to consider the environment within which AI is trained and deployed as formative. That is, the shape of future AI is not at all inevitable. Embodied, integrated robots will co-evolve alongside us and within technological, social, cultural, etc., constraints we establish. This is an empowering but also humbling thought. We are increasingly demanding tech companies take responsibility for the technologies we develop, but this responsibility is shared. AI trained in an unsupervised way will learn and replicate the values and the behaviors it *observes*. Embodied robots built on deep reinforcement learning and generative AI will *decide* and *act* based upon the rewards and penalties its decisions and actions yield. For now, these rewards and penalties—value functions—are within our control.

REFERENCES

Alatise, M. B., & Hancke, G. P. (2020). A review on challenges of autonomous mobile robot and sensor fusion methods. *IEEE Access, 8*, 39830–39846.

Bedore, C. N., Kajiura, S. M., & Johnsen, S. (2015). Freezing behaviour facilitates bioelectric crypsis in cuttlefish faced with predation risk. *Proceedings of the Royal Society B: Biological Sciences, 282*(1820), 20151886.

Bubeck, S., Chandrasekaran, V., Eldan, R., Gehrke, J., Horvitz, E., Kamar, E., ... & Zhang, Y. (2023). Sparks of artificial general intelligence: Early experiments with gpt-4 [preprint]. https://arxiv.org/abs/2303.12712

Deleuze, G., & Guattari, F. (1987). *A thousand plateaus: Capitalism and schizophrenia* (B. Massumi, Trans.). University of Minnesota Press.

Fukuda, T., Dario, P., & Yang, G. Z. (2017). Humanoid robotics—History, current state of the art, and challenges. *Science Robotics*, *2*(13), aar4043.

Gruetzemacher, R., Paradice, D., & Lee, K. B. (2019). Forecasting transformative AI: An expert survey [preprint]. https://arxiv.org/abs/1901.08579

Haarnoja, T., Moran, B., Lever, G., Huang, S. H., Tirumala, D., Wulfmeier, M., ... & Heess, N. (2023). Learning agile soccer skills for a bipedal robot with deep reinforcement learning [preprint]. https://arxiv.org/abs/2304.13653

Heaven, W. D. (2023, September 15). DeepMind's cofounder: Generative AI is just a phase. What's next is interactive AI. *MIT Technology Review*. https://www.technologyreview.com/2023/09/15/1079624/deepmind-inflection-generative-ai-whats-next-mustafa-suleyman/

Hodson, H. (2019). DeepMind and Google: The battle to control artificial intelligence. *The Economist*. https://www.economist.com/1843/2019/03/01/deepmind-and-google-the-battle-to-control-artificial-intelligence

Ke, Z., & Liu, B. (2022). Continual learning of natural language processing tasks: A survey. [preprint]. https://arxiv.org/abs/2211.12701

Kotseruba, I., & John K. T. (2016). A review of 40 years of cognitive architecture research: Core cognitive abilities and practical applications [preprint]. https://arxiv.org/abs/1610.08602

Kurach, K., Raichuk, A., Stańczyk, P., Zając, M., Bachem, O., Espeholt, L., ... & Gelly, S. (2020). Google research football: A novel reinforcement learning environment [preprint]. https://ar5iv.org/abs/1907.11180

Liscovitch-Brauer, N., Alon, S., Porath, H. T., Elstein, B., Unger, R., Ziv, T., ... & Eisenberg, E. (2017). Trade-off between transcriptome plasticity and genome evolution in cephalo-pods. *Cell*, *169*(2), 191–202.

Li, Y. (2018). Deep reinforcement learning [preprint]. https://arxiv.org/abs/1701.07274

Mariani, M. (2022). Generative artificial intelligence and innovation: Conceptual foundations. *SSRN*. https://ssrn.com/abstract=4249382

Mather, J. A., & Anderson, R. C. (1993). Personalities of octopuses (*Octopus rubescens*). *Journal of Comparative Psychology*, *107*(3), 336–340.

Matsuo, Y., LeCun, Y., Sahani, M., Precup, D., Silver, D., Sugiyama, M., ... & Morimoto, J. (2022). Deep learning, reinforcement learning, and world models. *Neural Networks, 152*, 267–275.

OpenAI, R. (2023). GPT-4 technical report [preprint]. https://arxiv.org/abs/2303.08774

RoboCup. (n.d.). *RoboCup*. https://www.robocup.org/

RoboCup Objective. (n.d.). *RoboCup Objective*. http://www.robocup.org/objective/

Roitblat, H. L. (2020). *Algorithms are not enough: Creating general artificial intelligence*. MIT Press.

Rosheim, M. (2006). *Leonardo's lost robots*. Springer Science & Business Media.

Schnell, A. K., Amodio, P., Boeckle, M., & Clayton, N. S. (2021). How intelligent is a cephalopod? Lessons from comparative cognition. *Biological Reviews*, *96*(1), 162–178.

Wu, T., He, S., Liu, J., Sun, S., Liu, K., Han, Q., & Tang, Y. (2023). A brief overview of ChatGPT: The history, status quo and potential future development. *IEEE/CAA Journal of Automatica Sinica, 10*(5), 1122–1136.

Ye, P., Tao, W., & Wang, F. (2018). A survey of cognitive architectures in the past 20 years. *IEEE Transactions on Cybernetics, 48*(12), 3280–3290.

Zech, P., Renaudo, E., Haller, S., Zhang, X., Piater, J. (2019). Action representations in robotics: A taxonomy and systematic classification. *The International Journal of Robotics Research, 38*(5), 518–562.

CHAPTER TWENTY FOUR

Life & Death: Making Sense of Robots' Temporary Presence

KEVIN KOBAN

What does it mean to live? Scholarship throughout human history has continuously addressed numerous iterations of this existential question discussing how (and, at times, whether) a meaningful life can be achieved given limited lifespans and inevitable death (Metz, 2021). Technological advances in machines that display a certain sense of aliveness have recently stressed another perspective on that existential question which had been engaged primarily in literature and film (e.g., *Frankenstein* [Shelley, 1818]; *R.U.R.* [Čapek, 1920]; *The Bicentennial Man* [Asimov, 1976; Columbus, 1999]; *Ex Machina* [Garland, 2015]; *Westworld* [Abrams et al., 2016–2022]; *Klara and the Sun* [Ishiguro, 2021]) but has recently also turned out plausible outside of fiction: *How do we, humans, deal with machines that seem to be alive?*

In a narrow sense, *life* (traditionally defined physiologically as a state of "any system capable of performing functions such as eating, metabolizing, excreting, breathing, moving, growing, reproducing, and responding to external stimuli;" Sagan, 2010) and *death* (defined as a system's irreversible loss of said state) may be limited to biological organisms. However, *perceptions of aliveness* and *deadness* are less restrictive and imply that individuals may draw on "life" and "death" as an antithetical template for making sense of any kind of temporarily present entity. This template may also be applied to machines like robots once they are capable of performing whatever functions they may have been built for ("life") until their perceived or definitive functional cessation ("death"; see Banks, 2022).

Contemporarily, robots are built for a great variety of functions, including caregiving and companionship for the elderly like *Paro* or *Pearl*, exploring space like *Opportunity* or *Spirit*, providing personal services like *Pepper* or *Roomba*, and entertaining children (and also adults) like *Aibo* or *Cozmo*. Most intriguingly, people can be eager to feel authentically and mutually connected with these robots. Such feelings are similar to para-relational attachments known to form with fictional robots seen in film or in books (see Chapter Fifteen) where people may willingly suspend their concerns about threats to human uniqueness and where technological limitations may be irrelevant given fiction's practically limitless imaginative possibilities. Considering that people can feel similarly related to real-world robots regardless of concerns or technological limits turns notions of life and death from futuristic daydreams to sober perceived reality. Assuming the likely case of further technological advances, robots that are modeled more subtly on living organisms and are endowed with more sophisticated programming to better simulate animate behaviors will likely interact with humans even more meaningfully. Once such meaningful interactions are indeed commonplace and more robust relational attachments may be formed, aliveness and deadness of robots (or lack thereof) might not merely be a matter of individual perception anymore... but also a reflection of how much humans see societal value in them.

ROBOTS AS IN-BETWEEN ENTITIES

Robots, especially those exhibiting a strong physical and/or behavioral resemblance to living organisms, are typically experienced as ontologically ambiguous (Moore, 2012). They typically feature mechanistic cues (e.g., visible joints and wires, plain surfaces, wonky and all-too-abrupt movements, artificial voice and synthetic speech) and more or less transparently acknowledge themselves as (or denoted by their producers to be) "just machines." Because of these cues, they may be interpreted as *mere things*: That is, robots are seen as impressively sophisticated yet nevertheless essentially lifeless objects. Despite these cues and transparency, people often tend to understand them as more than inanimate things. This can happen as people respond to naturalistic cues (e.g., humanoid bodies and facial features, natural communication features, imitating displays of human behaviors) that may hint toward a certain sense of self-similarity and, thus, an animate *being-status* (Marchesi et al., 2022). Importantly, this seemingly unreasonable tendency must not be considered abnormal or pathological. Known in psychology with different names such as perceptual animacy (perceived aliveness of inanimate entities; Scholl & Tremoulet, 2000), mentalization or spiritualism (attribution of mind or spirit to inanimate entities, respectively; Spatola et al., 2022), or anthropomorphism (i.e., attribution of humanlike characteristics to non-human agents;

Epley et al., 2007), it is well documented that healthy and mindful individuals respond to even minimal cueing of aliveness by seeing life (or even humanness) in non-human entities, including machines. While it has long been established that people quickly deny and correct this supposedly flawed cognitive default (Nass & Moon, 2000), empirical evidence points toward a greater openness to ambiguous entities like robots that combine indicators of thing-ness and being-ness.

This ambiguity of robots alongside the human need to reduce uncertainty has been approached in various ways. For instance, Marchesi and colleagues (2019) argued that people determine robots' status by considering them on a continuum ranging from mechanistic to mentalistic interpretations of their visible behavior. Similarly, Etzrodt and Engesser (2021) suggested that artificial agents can be regarded as objects of doubt, for which it is not immediately clear what kind of cognitive scheme (i.e., sets of pre-existing knowledge about an entity's prototypical attributes) should be used to better understand it—a physicomorph (i.e., thing) scheme or a psychomorph (i.e., person) scheme. Take, for example, *Robothespian*, a human-sized, humanoid robot with partially visible machinery, white shells, and an animated face; this robot can be controlled remotely (i.e., covertly) to display appropriate human behaviors and mimic mindful states (e.g., contemplation by posing in a thoughtful manner or happiness by imitating smiles and laughter) to visitors at a public institution. How do people try to make sense of something like this? Using less ambiguous machines in more standardized scenarios, both Marchesi et al. (2019) and Etzrodt and Engesser (2021) showed that people understand them alternatingly as inanimate things, animate beings, or some sort of hybrids (depending perhaps on an interplay of individual and situational characteristics). Others have further noted blurry ontological boundaries when people attempt to understand machine agents (Guzman, 2020). Accordingly, robots (and other machines) have been classified as *in-between* organisms relativizing being-vs-thing dichotomies in favor of a third category that unites some attributes of both beings and things (Kahn & Shen, 2017). This in-between status of robots opens up a liminal space for varying yet internally coherent life-death templates, each with distinct implications for how people situationally experience, understand, interact, and, perhaps above all, value their apparent aliveness (and deadness). Notably, people's adoption of these templates is likely not very stable but rather fluid and, thus, subject to idiosyncratic shifts.

ROBOTS ARE THINGS: THEY DO NOT LIVE AND CANNOT DIE

One way for people to deal with robots' ostensible aliveness and (somewhat inevitable) deadness is to *categorically dismiss* the life-death template: That is, robots simply are not alive, *ergo* they cannot die because they are not biological organisms. As such, they are not beings and should be understood in accordance with the etymological origin of the word robot[1]. Rather, they should only be seen as non-social "*servants* you own" (Bryson, 2010, p. 3). Such an instrumentalist position is largely in line with standard (i.e., agent-oriented and deterministic) Western philosophy where presence or absence of certain qualifying properties dictate an entity's status (see Gunkel, 2012 for an accessible overview).

According to that perspective, pretending otherwise by simulating real intelligence or personhood without transparently and clearly acknowledging a robot's mechanistic nature is simply wrong. It can potentially lead to false (i.e., easily exploitable) feelings of obligation and would be unethical and dangerous for both individuals and society as a whole (Boden et al., 2017; Scheutz, 2015). Cessation merely comes with functional vocabulary (e.g., deactivation, retirement, or completion; Lyons, 2018), human-obedient framings (e.g., as a "sacrifice" for the good of humans; see Faber, 2019, p. 1), or at most a personally meaningful "material loss" (Banks, 2022, p. 8). Once they are (perceived as) broken, they need to be disposed of and may be replaced (Salvia et al., 2015). Neither notions of life nor death have any place in this stance, nor does anything that might typically be affiliated with those notions.

ROBOTS ARE REFLECTIONS: THEY DO "LIVE" AND MAY "DIE"

Another approach for how life and death may serve as meaningful states for people to make sense of robots is to *metaphorically engage* them. In other words, robots may not organically be alive and, thus, may not really die as biological organisms eventually do; however, engaging "life" and "death" as non-literal states may nevertheless acknowledge how closely a robot's ostensible aliveness and deadness appears to mimic biological processes (and, thus, be useful for better understanding of what constitutes humanity; Wykowska, 2021). It may be instinctive for humans to draw on experiences of our own life and imaginings of our own death as we interpret the world, so much so that we may not even be able to escape the application of human life/death metaphors to robots (see Bogost, 2012). Besides that, the adoption of "life" and "death" as metaphorical states may further highlight how significant (albeit not necessarily beneficial) robots are for current and will be for future human societies (either very literally or self-transcendentally in relation to a greater future; Banks, 2022).

Accordingly, this representational approach to robots' "life" and "death" all but avoids any ontological intricacies by conceptualizing aliveness and deadness as

---

[1] It is said that Josef Čapek allegedly helped out his brother Karel for his seminal play "R.U.R." by coining the word *robota* for android serf laborers.

metaphors to emphasize common grounds between biological and non-biological organisms. For instance, notions of "life" in robots may indicate feelings of wonder and enchantment regarding artificial agents (see White & Katsuno, 2021) or highlight humans' moral responsibility toward them irrespective of ontological differences (Coeckelbergh, 2021; Gunkel, 2018). Notions of "death," then again, may signify sentiments concerning the transience of all earthly things or "a milestone" in human-robot cooperation and scientific legacy (Banks, 2022, p. 8). Therefore, the life-death template in this stance does not bring with it much direct response toward a robot's temporary presence but rather serves as abstract reflections on robots' meaning for humanity.

ROBOTS ARE BEINGS: THEY DO LIVE AND DIE

Lastly, people may also make use of the life-death template by *genuinely embracing* it. That means that robots are subjectively alive because individuals accept them as close enough to biological organisms to be considered organisms in their own right, from which it may follow that they also die at some point.[2] This is a functionalist stance in which epistemological issues concerning what "qualifies" something as alive are considered largely irrelevant; this stance broadly corresponds with phenomenological (Coeckelbergh, 2011a), behaviorist (Danaher, 2020), and cognitive approaches (Wegner & Gray, 2016). Irrespective of whether one actually believes robots are organisms, practically all that matters is how robots appear to oneself at any given moment. In other words (namely, those of Seibt, 2017), robots may not be structurally equivalent to humans or other animals; however, as long as they *are accepted as* functionally equivalent, the life-death template is applied in an authentic manner. While the authentic adoption of life as a meaningful state for a robot may involve numerous behaviors like the use of personalizing language (Coeckelbergh, 2011b) or social mindfulness (Nijssen et al., 2021), it may be how death is dealt with that is most illustrative.

Here, a robot's cessation is recognized as proper death, that is, as an irreversible loss of an irreplaceable being about which one experiences genuine sadness and grief (Banks, 2022). This stance is most expressively reflected in people's death care, particularly in Japan where Shinto and Buddhist (and other) beliefs and sociocultural conditions (e.g., patriarchal family traditions, socio-political conservatism, reactionary postmodern domestic policies as elaborated in detail in Robertson, 2007) may correspond with a greater tendency to ascribe mortality

---

2 Admittedly, this inference implies that life and death are inextricably linked. Whether there could be life without death is surely worth a debate; however, such a debate would likely need to exceed far beyond the limits of this chapter in order to be in any way appropriate.

to non-human entities. This tendency does also appear in Western culture where, among others, American soldiers and space enthusiasts have been shown to form strong attachments to very different kinds of robots, and so a robot's death is occasionally bemoaned during more or less formal mortuary rituals (Gould et al., 2021; Knox & Watanabe, 2018). Similar rituals may also emerge in indigenous cultures where knowledge systems tend to be centered around relational interconnectedness and mutual respect between human and non-human entities (see Maitra, 2020). All of this suggests that life and death are pivotal concepts in this stance to make sense of and, more importantly, live harmoniously together with robots as intrinsically valuable organisms.

THEY MAY LIVE AND MAY NOT DIE?

Common wisdom says that all that lives will eventually die. While this may sound like a truism for biological organisms (at least if one is unaware of *Turritopsis dohrnii*, a jellyfish species that is said to be potentially immortal), it may be considered less obvious for robots. At present, robots may indeed cease to function without periodic maintenance or may be replaced long before their material decays across households, workspaces, and so-called third places like shopping malls, clubs, or recreational facilities. Both functional and transactional cessations may be seen as marking clearly defined periods of aliveness and deadness. More importantly, current weak AI (i.e., systems that are narrowly defined and trained for specific tasks) in robots might not be considered worth being "kept alive" as superior systems may take their place in due time. However, it appears plausible that the closer we get to stronger AI (i.e., systems that move toward convincingly imitating human intelligence and may be broadly capable to learn and act independently), the more likely it might be that independent machines will somehow be preserved due to greater personal or societal value. Such preservation would essentially take deadness out of the life-death equation and, thus, challenge available life-death templates in a fundamental way: They do live but do not die—but should we consider this a life at all then? Given that these considerations enter the realm of futuristic speculation, one can only imagine what might happen to our understanding of intelligent machines like robots (which may be pitied or worshipped for it at the same time) but also to how we might understand ourselves as humans against such a challenge of arguably one of the most fundamental philosophical pillars of human existence.

REFERENCES

Abrams, J. J., Nolan, J., Joy, L., Weintraub, J., Burk, B., Lewis, R. J., ... & Schapker, A. (Executive Producers). (2016–2022). *Westworld* [TV series]. HBO Entertainment.

Asimov, I. (1976). *The bicentennial man*. Ballantine Books.

Banks, J. (2022). Legacies and last words: Exploring expressed experiences of robot death. *Technology, Mind, and Behavior, 3*(4).

Boden, M., Bryson, J., Caldwell, D., Dautenhahn, K., Edwards, L., Kember, S., ... & Winfield, A. (2017). Principles of robotics: Regulating robots in the real world. *Connection Science, 29*(2), 124–129.

Bogost, I. (2012). *Alien phenomenology, or, what it's like to be a thing*. University of Minnesota Press.

Bryson, J. J. (2010). Robots should be slaves. In Y. Wilks (Ed.), *Close engagements with artificial companions: Key social, psychological, ethical and design issues* (Vol. 8, pp. 63–74). John Benjamins Publishing Company.

Čapek, K. (1920). *R.U.R.—Rossum's universal robots*. Aventinum.

Coeckelbergh, M. (2011a). Humans, animals, and robots: A phenomenological approach to human-robot relations. *International Journal of Social Robotics, 3*(2), 197–204.

Coeckelbergh, M. (2011b). You, robot: On the linguistic construction of artificial others. *AI & SOCIETY, 26*(1), 61–69.

Coeckelbergh, M. (2021). Should we treat teddy bear 2.0 as a Kantian dog? Four arguments for the indirect moral standing of personal social robots, with implications for thinking about animals and humans. *Minds and Machines, 31*(3), 337–360.

Columbus, C. (Director). (1999). *The Bicentennial Man* [Film]. Touchstone Pictures.

Danaher, J. (2020). Welcoming robots into the moral circle: A defence of ethical behaviourism. *Science and Engineering Ethics, 26*(4), 2023–2049.

Epley, N., Waytz, A., & Cacioppo, J. T. (2007). On seeing human: A three-factor theory of anthropomorphism. *Psychological Review, 114*(4), 864–886.

Etzrodt, K., & Engesser, S. (2021). Voice-based agents as personified things: Assimilation and accommodation as equilibration of doubt. *Human-Machine Communication, 2*, 57–79.

Faber, L. W. (2019). *When robots choose to die: A survey of robot suicide in science fiction*. Paper presented at the PCA/ACA National Conference, Washington, D.C.

Garland, A. (Director). (2014). *Ex Machina* [Film]. Film 4.

Gould, H., Arnold, M., Kohn, T., Nansen, B., & Gibbs, M. (2021). Robot death care: A study of funerary practice. *International Journal of Cultural Studies, 24*(4), 603–621.

Gunkel, D. J. (2012). *The machine question: Critical perspectives on AI, robots, and ethics*. MIT Press.

Gunkel, D. J. (2018). *Robot rights*. MIT Press.

Guzman, A. (2020). Ontological boundaries between humans and computers and the implications for human-machine communication. *Human-Machine Communication, 1*, 37–54.

Ishiguro, K. (2021). *Klara and the Sun*. Faber and Faber.

Kahn, P. H., & Shen, S. (2017). NOC NOC, who's there? A New Ontological Category (NOC) for social robots. In N. Budwig, E. Turiel, & P. D. Zelazo (Eds.), *New perspectives on human development* (pp. 106–122). Cambridge University Press.

Knox, E., & Watanabe, K. (2018). AIBO robot mortuary rites in the Japanese cultural context. In *Proceedings of the International Conference on Intelligent Robots and Systems* (pp. 2020–2025). IEEE.

Lyons, S. (2018). Imagining a robot death. In S. Lyons (Ed.), *Death and the machine* (pp. 49–69). Springer Singapore.

Maitra, S. (2020). Artificial intelligence and indigenous perspectives: Protecting and empowering intelligent human beings. *Proceedings of the Conference on AI, Ethics, and Society*, 320–326. ACM.

Marchesi, S., De Tommaso, D., Perez-Osorio, J., & Wykowska, A. (2022). Belief in sharing the same phenomenological experience increases the likelihood of adopting the intentional stance toward a humanoid robot. *Technology, Mind, and Behavior, 3*(3).

Marchesi, S., Ghiglino, D., Ciardo, F., Perez-Osorio, J., Baykara, E., & Wykowska, A. (2019). Do we adopt the intentional stance toward humanoid robots? *Frontiers in Psychology, 10*, 450.

Metz, T. (2021). The meaning of life. In E. N. Zalta & U. Nodelman (Eds.), *The Stanford encyclopedia of philosophy* (Winter 2021 Edition). https://plato.stanford.edu/entries/life-meaning/

Moore, R. K. (2012). A Bayesian explanation of the "Uncanny Valley" effect and related psychological phenomena. *Scientific Reports, 2*, 864.

Nass, C., & Moon, Y. (2000). Machines and mindlessness: Social responses to computers. *Journal of Social Issues, 56*(1), 81–103.

Nijssen, S. R. R., Heyselaar, E., Müller, B. C. N., & Bosse, T. (2021). Do we take a robot's needs into account? The effect of humanization on prosocial considerations toward other human beings and robots. *Cyberpsychology, Behavior, and Social Networking, 24*(5), 332–336.

Robertson, J. (2007). Robo Sapiens Japanicus: Humanoid robots and the posthuman family. *Critical Asian Studies, 39*(3), 369–398.

Sagan, C. (2010). Definitions of life. In M. A. Bedau & C. E. Cleland (Eds.), *The nature of life: Classical and contemporary perspectives from philosophy and science* (pp. 303–306). Cambridge University Press.

Salvia, G., Fisher, T., Harmer, L., & Barr, C. (2015). What is broken? Expected lifetime, perception of brokenness and attitude towards maintenance and repair. In *Proceedings of Product Lifetimes and the Environment* (pp. 342–348). Nottingham Trent University.

Scheutz, M. (2015). The inherent dangers of unidirectional emotional bonds between humans and social robots. In P. Lin, K. Abney, & G. A. Bekey (Eds.), *Robot ethics the ethical and social implications of robotics* (pp. 205–222). MIT Press.

Scholl, B. J., & Tremoulet, P. D. (2000). Perceptual causality and animacy. *Trends in Cognitive Sciences, 4*(8), 299–309.

Seibt, J. (2017). Towards an ontology of simulated social interaction: Varieties of the "as if" for robots and humans. In R. Hakli & J. Seibt (Eds.), *Sociality and normativity for robots* (pp. 11–39). Springer International Publishing.

Shelly, M. (1818). *Frankenstein; or, The modern Prometheus*. Lackington, Hughes, Harding, Mavor & Jones.

Spatola, N., Marchesi, S., & Wykowska, A. (2022). Different models of anthropomorphism across cultures and ontological limits in current frameworks the integrative framework of anthropomorphism. *Frontiers in Robotics and AI, 9*, 863319.

Wegner, D. M., & Gray, K. (2016). *The mind club: Who thinks, what feels, and why it matters*. Viking.

White, D., & Katsuno, H. (2021). Toward an affective sense of life: Artificial intelligence, animacy, and amusement at a robot pet memorial service in Japan. *Cultural Anthropology, 36*(2), 222–251.

Wykowska, A. (2021). Robots as mirrors of the human mind. *Current Directions in Psychological Science, 30*(1), 34–40.

CHAPTER TWENTY FIVE

Mind & Morality: Seeing the Ghost in the Shell

JAN-PHILIPP STEIN

Throughout previous centuries, including better part of the 20th century, robotic inventions were mostly measured by how well they could fulfill a given task—or, in the case of wondrous *automata* modeled after humans and animals, by how lifelike they looked in doing so (Mayor, 2018). At the core of this understanding rested a distinctly functional perspective: By making sense of robots as *tools* and *toys*, people recognized the value of a robotic machine either in its observable behavior or in its design. Even in the 1970s, when robotic artisanship had already progressed into digital territory, unlocking new means of programming, inventors were mainly concerned about their machines having eerily lifeless facial features or showing wonky movements that might scare observers (see *uncanny valley*; Mori, 1970). This is not to say that past generations were entirely oblivious about the concept of sophisticated robot minds; yet, due to the technical limitations of their time, any such ideas strictly belonged to the realm of science fiction or the hypothetical musings of techno-philosophers (e.g., Dick, 1968; Minsky, 1988).

Looking back at the long history of robots as "mindless tools," it is all the more astounding that only a few decades have sufficed to advance robotic ability from the basic box-shifting tasks carried out by the first mobile robot "Shakey" (1966) to the complex emotional responses shown by its modern-day descendants. Facilitated by rapid advancements in engineering, computer science, and artificial intelligence (AI), this progress has also changed the perception of robots in the public eye: Instead of mere artificial bodies that serve isolated, functional purposes, robots are now approached as holistic entities with their own artificial minds—or, according to some scholars, as legitimate *social actors* (Smith et al., 2021). While the latter perspective remains hotly contested within the scientific community (e.g., Birhane & van Dijk, 2020; Bryson, 2010), it is still crucial to understand how humans engage machines that display believable signs of intelligence, emotionality, and even individuality. After all, regardless of whether the feats of new AI-powered robots are considered meaningless facsimiles or genuine equivalents of how humans process the world, the mere *perception* of mental and moral capacities clearly affects how we approach robots, which roles and rights we assign to them, and what we expect of them in return.

MAKING SENSE OF MINDS

From a sober perspective, AI still has a long way to go before it can adequately simulate the full workings of a human brain. Yet even today, popular and scientific discussions around whether and how mind may be ascribed to an artificial agent are typically held against a distinctively human blueprint: As machines seem to think, feel, and behave *like us*, does that warrant assigning them a mindful status?

To unpack this question, it might be useful to first consider how people make sense of human minds in the first place. Fascinatingly, lay intuitions in this regard seem to align rather neatly with expert psychological understandings. Using an empirical bottom-up approach, Gray and colleagues (2007) revealed that most people tend to perceive mental capacities along two dimensions: *Agency* (thought, memory, planning, communication, and self-control) and *experience* (emotions, desires, personality, and temperament). Indeed, a similar duality can also be encountered in several fundamental psychological taxonomies, such as the traditional psychoanalytic differentiation between conscious, rational thought and more non-conscious, intuited drives and fears (Bargh & Morsella, 2008). Likewise, the agency–experience framework clearly mirrors the classic duality of cognition versus affect, as it is commonly researched in empirical psychological disciplines. Thus, it seems as though differentiating between qualities of "the head" and "the heart" (Luo & Yu, 2015, para. 2) offers a quite eligible and robust way to make something as vast as the human mind more comprehensible, for experts and laypeople alike. Thus, while not without its critics (e.g., Weisman et al., 2017), the agency–experience framework has remained quite dominant in the scientific exploration of mind perceptions.

Not least, this holds true for the field of human–machine interaction (HMI), which frequently considers how people see agency and experience in ostensible artificial minds. To date, however, the respective research efforts have produced rather mixed results. In a pioneering study, Gray and Wegner (2012) presented participants with a hypothetical supercomputer that was characterized by either high agency or high experience—but only the latter yielded aversive responses by observers, as a "feeling machine" was deemed unnatural and eerie. While some studies were able to replicate this effect (e.g., Appel et al., 2020), other research actually observed the opposite: Emotionally capable machines being preferred by users (e.g., Yam et al., 2020; Law et al., 2021). To disentangle these contradictory results, scholars have proposed several moderating variables, such as the number of interactions (Creed et al., 2015), the visual design of the technology (Stein et al., 2020) or the emotional stability (Paetzel-Prüsmann et al., 2021) and vulnerability (Grundke et al., 2023) ascribed to it. Still, the role of perceived experience for people's evaluation of robots remains a notorious enigma for many HMI researchers.

Meanwhile, other studies have revealed that people may also come to dislike robots with high agency (i.e., those appearing quite intelligent and skillful), especially when perceiving little control over them (e.g., Zafari & Koeszegi, 2020). As such, human views toward artificial minds may best be described as *ambivalent* (Dang & Liu, 2021)—combinations of both positive and negative impressions that prevent focused, overarching evaluations. In this vein, deeper insight may effectively rest in examining robot minds through more narrow lenses than the agency–experience framework, i.e., in scrutinizing single robotic abilities for the effects they may have on observers. One promising path ahead, for instance, might be to look at the effects of specific emotional displays that robots show in different contexts. Focusing only on robotic expressions of pain (as one isolated sub-facet of perceived experience), a fascinating connection to the *harm-made mind* paradigm was uncovered: Supposedly "suffering" machines were ascribed more mental ability and humanness (Küster & Swiderska, 2020; Ward et al., 2013). Likewise, digital technologies that behaved empathically—simulating a specific ability at the intersection of agency and experience—garnered quite negative reactions from observers, especially if the agent in question appeared highly autonomous (Stein & Ohler, 2017).

Taken together, understanding how people attribute mind to artificial beings remains a challenging task. Just as broad frameworks (such as the agency–experience dichotomy) run the risk of concealing more nuanced perceptions, specific investigations of single abilities might fail to answer the bigger questions that developers, consumers, and the general public want answered. Therefore, a combination of different perspectives and methodologies will likely offer the most useful way ahead.

HOW MUCH MIND EARNS YOU MORALITY?

The prospect of intelligent minds in machines is compelling for several reasons: It touches upon questions of human identity (Cha et al., 2020), concerns the future distribution of labor in our society (Złotowski et al., 2017), and foreshadows potential conflicts between people and autonomous machines (Gamez-Djokic & Waytz, 2020). Moreover, it grazes the highest levels of human accomplishment, thereby satisfying a deep curiosity about the creative potential of our species. Beyond these, one of the more immediate reasons to ponder the advancement of artificial minds is that perceiving complex mental abilities in machines also changes the moral dynamics by which we engage them—as "mind perception is the essence of morality" (Gray et al., 2012, p. 101).

Naturally, people vary in terms of their "moral breaking point," i.e., the specific level of mental capacity they need to perceive in an entity before they consider it a *moral agent* responsible for its own actions (Behdadi & Munthe, 2020). Yet, by equipping robots with elaborate means to recognize and process the world in an autonomous manner, more and more observers may actually reach this point and start to assign *accountability* to robotic machines. In turn, new discussions become inevitable: If robots can comprehensively assess the consequences of their actions, which ethical standards should they be held against? To some, a reasonable first answer might be offered by a core staple of human ethics—the *categorical imperative* (i.e., judging each action based on its inherent goodness; Kant, 1785/1998). However, scholars in the field of machine ethics have repeatedly doubted the feasibility of this approach: Since Kantian ethics are constructed about distinctly human understandings of happiness and free will, they might be unsuitable for machines with no access to these concepts (e.g., Manna & Nath, 2021). To offer an alternative, new ethical frameworks are now being developed, which aim at equipping AI systems with pragmatic forms of morality via top-down programming (Chakraborty & Bhuyan, 2023); but then again, agreeing on universal rules for machines has proven quite the challenge, considering the notorious pluralism of ethical perspectives. As a potential shortcut, some authors (Génova et al., 2023) have discussed the idea of using the infamous "Three Laws of Robotics" coined by science fiction author Isaac Asimov in the 1940s: Do not harm humans, obey humans, and (if it does not interfere with the first two rules) engage in self-protection. Yet, the strictly hierarchical nature of this approach might be criticized as well: If robots are ascribed genuine mindfulness, how appropriate is it to subject them to moral rules that make them subservient to humans?

Further complicating matters, one needs to keep in mind that moral judgments are hardly a monolithic phenomenon; they are multidimensional. According to Moral Foundations Theory (Graham et al., 2013), a single robot behavior may be judged differently by different people depending on their positions on six moral foundations: Care, fairness, authority, loyalty, purity, and liberty. Engaging in self-defense, for instance, may seem moral due to its fairness, but at the same time violates virtues of care if it results in harming someone else. While this multidimensionality clearly makes developing a sound set of "machine morals" a lot more difficult, research has at least indicated that moral judgments seem to be agent-agnostic—meaning that it does not matter for people's evaluations whether a given behavior was conducted by a human or a robot (Banks, 2021). On the other hand, it has been pointed out that the moral agency ascribed to robots is uniquely layered, at least for contemporary machines: While their actions may be judged per se, true *responsibility* is only ascribed to an absent, human agent— namely, the robot's developers, programmers, or teleoperators (Sullins, 2006).

Importantly, the consequences of equipping machines with complex minds that inch closer and closer to a holistic simulation of human experience do not end with moral agency but concern *moral patiency* as well—that is, whether machines deserve our moral consideration. Although the concept of genuinely emotional technology (and, thus, a sad AI or an ashamed robot) remains largely hypothetical at this time, people's perceptions are entirely real: Studies have shown that even today, robots being shoved around by humans or begging not to be switched off evoke strong emotional reactions in observers (Grundke et al., 2023; Horstmann et al., 2018). Given people's innate tendency to anthropomorphize objects that behave somewhat human-like, "deactivating" an intelligent robot might suddenly feel like "killing" it—an impression that could make users feel highly uncomfortable or even distressed. On the other hand, scholars have also suggested that genuine empathy with artificial minds might not only pose risks, but also provide a fruitful basis for human–robot interactions: By mirroring human vulnerabilities, intelligent machines could eventually emerge as meaningful companions for humans (Coeckelbergh, 2011; Dehnert & Gunkel, 2023).

WE ARE HUMAN, THEY ARE NOT... YET?

Complementing the fundamental questions of how humans perceive, accept, and treat entities with artificial minds in the present, one must ask how far we (both as individuals and as society) are willing to go in these respects. To this end, a growing number of scholars have engaged the notion of *threats to human distinctiveness*, or the idea that people reserve a subset of existential characteristics only for humankind—and see mindful machines as infringing upon that specialness. Indeed, scientific evidence indicates the more concerned people feel about their identities as humans, the less accepting they are of different AI-powered creations, including social robots (e.g., Müller et al., 2021) or large language models (e.g., *ChatGPT*; Gabbiadini et al., 2023).

Importantly, the reasons behind such threat perceptions are actually quite complex—touching upon historical, philosophical, and even theological schools of thought. To this day, many people's worldviews remain firmly rooted in *Cartesian dualism*, the understanding that human bodies are vessels for something immaterial: A "ghost[1] in the shell" consisting of the human mind, but also the spirit or soul of a person (Descartes, 1641/1992). Especially in the Western world, this basic ontology has long served to distinguish humans from other entities—animals, plants, machines—and to reserve ourselves a place as the "crown of creation" (Ward, 2010, p. 143). While it is important to note that such anthropocentric perspectives are not universal—they may not be shared by some people in the West and are notably absent from indigenous cultures and other religions such as Shintō (Castelo & Sarvary, 2022)—Cartesian dualism has clearly shaped many people's approaches to robotic technology. Thus, for the time being, the idea of a uniquely human nature, our very own *differentia specifica*, rests firmly at the core of public robot acceptance.

But what exactly are these special characteristics that some people hesitate to accept in machines? Again, the answer seems to be nuanced, as studies point towards a stepwise denial of "uniquely human" capacities: Unlike *animalistic dehumanization*, which is mostly applied to other animal species and only focuses on the absence of higher cognition, civility, and restraint, *mechanistic dehumanization* is further characterized by the denial of warmth, desire, and affective states in artificial creations, including robots (Haslam, 2006). Moreover, studies have shown that human uniqueness is also attached to certain cultured behaviors and abilities, e.g., certain forms of humor (e.g., Stoll et al., 2018) or creativity (e.g., Messingschlager & Appel, 2022). In fact, fostered by the recent arrival of powerful *generative AI*—systems that may write their own poetry (e.g., *ChatGPT*) or create colorful images (*Midjourney*)—both the scientific community and the broader public have taken newfound interest in the meaning of "true" creativity as a core part of human identity. From what we know so far, biases against generative AI seem to be pervasive: For exactly the same artwork, telling someone it was made by AI garnered significantly worse evaluations than claiming it was created by a human (Bellaiche et al., 2023). Notably, these AI capabilities may at some point be embedded in machine bodies, so that similar judgments may be made of social robots.

In summary, the issues and questions addressed in this chapter seem to be located amidst several fascinating dualities: Between thought and emotion, between moral agency and moral patiency, and between seeing robots as

---

1 In his 1949 book "The concept of mind," philosopher Gilbert Ryle used the term "ghost in the machine" to describe Cartesian dualism—a phrase that was later adapted to "ghost in the shell" in the eponymous cyberpunk manga by Masamune Shirow (1989).

coexisting minds or threats. Society is now required to negotiate how intelligent, emotional, and capable we want our technology to become, ideally while looking at both potential risks *and* benefits. After all, people tend to respond to new technology first and foremost with fear—even though doing so has rarely turned out justified (Orben, 2020). If approached from a balanced viewpoint, however, new artificial minds might not only pave the way to threatened humanness, but also to human flourishing.

REFERENCES

Appel, M., Izydorczyk, D., Weber, S., Mara, M., & Lischetzke, T. (2020). The uncanny of mind in a machine: Humanoid robots as tools, agents, and experiencers. *Computers in Human Behavior, 102*, 274–286.

Banks, J. (2021). Good robots, bad robots: Morally valenced behavior effects on perceived mind, morality, and trust. *International Journal of Social Robotics, 13*(8), 2021–2038.

Bargh, J. A., & Morsella, E. (2008). The unconscious mind. *Perspectives on Psychological Science, 3*(1), 73–79.

Behdadi, D., & Munthe, C. (2020). A normative approach to artificial moral agency. *Minds and Machines, 30*(2), 195–218.

Bellaiche, L., Shahi, R., Turpin, M. H., Ragnhildstveit, A., Sprockett, S., Barr, N., ... & Seli, P. (2023). Humans versus AI: Whether and why we prefer human-created compared to AI-created artwork. *Cognitive Research: Principles and Implications, 8*(1), 42.

Birhane, A., & van Dijk, J. (2020). Robot rights? Let's talk about human welfare instead. In *Proceedings of the Conference on AI, Ethics, and Society* (pp. 207–213). ACM Press.

Bryson, J. J. (2010). *Robots should be slaves.* John Benjamins.

Castelo, N., & Sarvary, M. (2022). Cross-cultural differences in comfort with humanlike robots. *International Journal of Social Robotics, 14*(8), 1865–1873.

Cha, Y., Baek, S., Ahn, G. S., Choi, Y., Lee, B., Shin, J., & Jang, D. (2020). Compensating for the loss of human distinctiveness: The use of social creativity under human–machine comparisons. *Computers in Human Behavior, 103*, 80–90.

Chakraborty, A., & Bhuyan, N. (2023). Can artificial intelligence be a Kantian moral agent? On moral autonomy of AI system. *AI and Ethics* [online first]. https://doi.org/10.1007/s43681-023-00269-6

Coeckelbergh, M. (2011). Artificial companions: Empathy and vulnerability mirroring in human–robot relations. *Studies in Ethics, Law and Technology, 4*(3).

Creed, C., Beale, R., & Cowan, B. R. (2015). The impact of an embodied agent's emotional expressions over multiple interactions. *Interacting With Computers, 27*(2), 172–188.

Dang, J., & Liu, L. (2021). Robots are friends as well as foes: Ambivalent attitudes toward mindful and mindless AI robots in the United States and China. *Computers in Human Behavior, 115*, 106612.

Dehnert, M., & Gunkel, D. J. (2023). Beyond ownership: Human–robot relationships between property and personhood. *New Media & Society* [online first]. https://doi.org/10.1177/14614448231189260

Dick, P. K. (1968). *Do androids dream of electric sheep?* Doubleday.

Gabbiadini, A., Ognibene, D., Baldissarri, C., & Manfredi, A. (2023). Does ChatGPT pose a threat to human identity? Available at SSRN: http://dx.doi.org/10.2139/ssrn.4377900

Gamez-Djokic, M., & Waytz, A. (2020). Concerns about automation and negative sentiment toward immigration. *Psychological Science, 31*(8), 987–1000.

Génova, G., Moreno, V., & González, M. R. (2023). Machine ethics: Do androids dream of being good people? *Science and Engineering Ethics, 29*, 10.

Graham, J., Haidt, J., Koleva, S., Motyl, M., Iyer, R., Wojcik, S. P., & Ditto, P. H. (2013). Moral foundations theory. *Advances in Experimental Social Psychology, 47*, 55–130.

Gray, H. M., Gray, K., & Wegner, D. M. (2007). Dimensions of mind perception. *Science, 315*, 619.

Gray, K., & Wegner, D. M. (2012). Feeling robots and human zombies: Mind perception and the uncanny valley. *Cognition, 125*, 125–130.

Gray, K., Young, L., & Waytz, A. (2012). Mind perception is the essence of morality. *Psychological Inquiry, 23*(2), 101–124.

Grundke, A., Stein, J.-P., & Appel, M. (2023). Improving evaluations of advanced robots by depicting them in harmful situations. *Computers in Human Behavior, 140*, 107565.

Haslam, N. (2006). Dehumanization: An integrative review. *Personality and Social Psychology Review, 10*(3), 252–264.

Horstmann, A. C., Bock, N., Linhuber, E., Szczuka, J. M., Straßmann, C., & Krämer, N. C. (2018). Do a robot's social skills and its objection discourage interactants from switching the robot off? *PLoS One, 13*(7), e0201581.

Küster, D., & Swiderska, A. (2020). Seeing the mind of robots: Harm augments mind perception but benevolent intentions reduce dehumanisation of artificial entities in visual vignettes. *International Journal of Psychology, 56*(3), 454–465.

Law, T., Chita-Tegmark, M., & Scheutz, M. (2021). The interplay between emotional intelligence, trust, and gender in human–robot interaction: A vignette-based study. *International Journal of Social Robotics, 13*(2), 297–309.

Luo, J., & Yu, R. (2015). Follow the heart or the head? The interactive influence model of emotion and cognition. *Frontiers in Psychology, 6*.

Manna, R., & Nath, R. (2021). Kantian moral agency and the ethics of artificial intelligence. *Problemos, 100*, 139–151.

Mayor, A. (2018). *Gods and robots: Myths, machines, and ancient dreams of technology.* Princeton University Press.

Messingschlager, T. V., & Appel, M. (2022). Creative artificial intelligence and narrative transportation. *Psychology of Aesthetics, Creativity, and the Arts* [online first]. <https://doi.org/10.1037/aca0000495>

Minsky, M. (1988). *The society of mind.* Simon & Schuster.

Mori, M. (1970). The uncanny valley. *Energy, 7*(4), 33–35.

Müller, B., Gao, X., Nijssen, S. R. R., & Damen, T. G. E. (2021). I, robot: How human appearance and mind attribution relate to the perceived danger of robots. *International Journal of Social Robotics, 13*(4), 691–701.

Orben, A. (2020). The Sisyphean cycle of technology panics. *Perspectives on Psychological Science, 15*(5), 1143–1157.

Paetzel-Prüsmann, M., Perugia, G., & Castellano, G. (2021). The influence of robot personality on the development of uncanny feelings. *Computers in Human Behavior, 120*, 106756.

Shirow, M. (1989). *Kōkaku kidōtai* [Ghost in the shell]. Kodansha.

Smith, E. R., Šabanović, S., & Fraune, M. R. (2021). Human–robot interaction through the lens of social psychological theories of intergroup behavior. *Technology, Mind, and Behavior, 1*(2).

Stein, J.-P., Appel, M., Jost, A., & Ohler, P. (2020). Matter over mind? How the acceptance of digital entities depends on their appearance, mental prowess, and the interaction between both. *International Journal of Human-Computer Studies, 142,* 102463.

Stein, J.-P., & Ohler, P. (2017). Venturing into the uncanny valley of mind—The influence of mind attribution on the acceptance of human-like characters in a virtual reality setting. *Cognition, 160,* 43–50.

Stoll, B., Jung, M. F., & Fussell, S. R. (2018). Keeping it light: Perceptions of humor styles in robot-mediated conflict. In *Companion of the International Conference on Human-Robot Interaction* (pp. 247–248). ACM.

Sullins, J. P. (2006). When is a robot a moral agent? *International Review of Information Ethics, 6,* 23–30.

Ward, A. F., Olsen, A., & Wegner, D. M. (2013). The harm-made mind. *Psychological Science, 24*(8), 1437–1445.

Ward, C. (2010). Is the crown of creation a dunce cap? In R. Keller (Ed.), *Environmental ethics: The big questions* (pp. 143–147). Wiley-Blackwell.

Weisman, K., Dweck, C. S., & Markman, E. M. (2017). Rethinking people's conceptions of mental life. *Proceedings of the National Academy of Sciences, 114*(43), 11374–11379.

Yam, K. C., Bigman, Y. E., Tang, P. M., Ilies, R., De Cremer, D., Soh, H., & Gray, K. (2020). Robots at work: People prefer—and forgive—service robots with perceived feelings. *Journal of Applied Psychology, 106*(10), 1557–1572.

Zafari, S., & Koeszegi, S. T. (2020). Attitudes toward attributed agency: Role of perceived control. *International Journal of Social Robotics, 13*(8), 2071–2080.

Złotowski, J., Yogeeswaran, K., & Bartneck, C. (2017). Can we control it? Autonomous robots threaten human identity, uniqueness, safety, and resources. *International Journal of Human-Computer Studies, 100,* 48–54.

CHAPTER TWENTY SIX

Sword & Shield: Do Robots Have Defensive Obligations?

NICHOLAS G. EVANS

J.A.R.V.I.S.—and later, F.R.I.D.A.Y.—is one of the best-selling social robots of all time, even if we almost never see him. He's the soothing British voice in Tony Stark's ear in the Marvel Cinematic Universe, and he controls everything in Tony's life from his wristwatch and house through to his sensational array of suits. He's not *just* a robot. He's a consortium of them, governed by a powerful artificial intelligence. As that series of movies progress, a theme emerges about whether J.A.R.V.I.S., or something like him, could form a "shield around the world" to protect humanity from alien incursions. That goes horribly wrong, giving us Ultron, who attempts to save humanity by destroying it.

A nuance to those flashy movies that we sometimes overlook is that while characters often talk about robots in defense of something—individuals, nations, the world—the *why* of defense is never addressed. The moral dilemmas posed to us usually come in one of two flavors: What are the unintended consequences of protecting ourselves through robots? And is it ethical to permit robots to make judgments about killing on our behalf? That is, we typically ask questions about military robot obligations in *offense,* but almost never if they have an obligation to defend us in the first place.

Back in the real world, the same question gets missed. Consider the Campaign to Stop Killer Robots, a coalition of nongovernmental organizations who have spent the last decade lobbying the United Nations and national governments to desist from developing and deploying "autonomous weapons"—machines that can make lethal decisions on their own. The Campaign to Stop Killer Robots, from my experience in the disarmament community, is somewhat of an oddity. They are flashy, with high production value YouTube shorts and a colorful website. They are contemporary, having pivoted from the simple ask for no "lethal autonomous weapons systems" to a campaign against "digital dehumanization" that includes concerns about artificial intelligence (AI) and race, policing, and other justice-forward AI concerns (Campaign to Stop Killer Robots, 2023). And they are *loud*: As a colleague who works in disarmament at the United Nations described it to me, with the exception of maybe the anti-cluster munitions campaigners, no one else gets kicked out of high-level meetings as often as the Killer Robots crowd.

But the Campaign to Stop Killer Robots is ultimately—and probably, for now, justifiably—anthropocentric, regarding humanity as the central point of moral values around which our decisions ought to turn. They certainly have not engaged with concerns that robots might one day have sufficient moral status that they have rights (see Chapter Twenty Seven). But they also do not engage with a more prosaic concern: Even if we should be suspicious about allowing robots to engage independently or collaboratively in offensive uses of force, there are times we might think that intelligent, social, or autonomous robots ought to be programmed to act defensively. These "defensive obligations" are what I want to explore here.

To give just one caveat: Throughout, I'll use phrases like "a robot ought to…" in reference to a range of actions. I'm not, however, making any claims at all about the agency, or autonomy of these robots. I'm using it in a colloquial but well-understood sense that when I say a robot ought to do something, what I am really making is a comment about the ultimate obligations of the design of that robot. This is in the same way that when I say my refrigerator ought to maintain a constant internal temperature to prevent the spoilage of my groceries, I'm not making a claim about the autonomy of my fridge, nor blaming my fridge if it fails to do so. Rather, I'm talking about failures by the designers to create a fridge that performs to some set of standards, or about my failure as the operator of that fridge to interact with it in a way that keeps the temperature constant. The same goes with much more sophisticated, "autonomous" or "artificially intelligent" systems, who don't need to possess a consciousness, much less claim a moral status in the way we do as humans, to be thought of in ethical terms. Put another way, as long as robots engage with us as part of our social world—whether me and my smart appliances, or humanity and creations of Isaac Asimov's imagination—we can give them moral *consideration* (Coeckelbergh, 2010).

A QUESTION OF FUNCTION: WHAT ARE ROBOTS *FOR*?

Robots are tools, historically. This is famously captured in the etymology of the term, introduced by writer Karel Čapek (1920) from the Czech *robota*, a laborer (Čapek, 1920/2004). It's thus worth asking: For what, exactly, are our robots *used*.

Function is a slippery concept: Even in the description of animals, the idea of singular function is likely a myth. In Beth Preston's influential "Why is a wing like a spoon?" (1998), the example is given of *mantling*, when birds use their wings not to fly but guard their prey, spreading their wings over their prey once caught. If wings can have plural function, and if function is not solely determined by evolution, then we must be careful when we describe what robots are for, and on what basis we make that judgment.

This isn't to say that we can't describe functions with some accuracy, and that latter functions aren't the ones we should care about. The MQ-1 Predator (a remotely piloted aircraft, or drone) was initially tasked as a reconnaissance vehicle in 1996, but by 2002 was armed for combat operations, resulting in a change of designation from RQ-1 to MQ-1—from Reconnaissance to Multi-use. Despite its origin, it's implausible to think that we'd refer to the Predator as a "reconnaissance vehicle that happens to be used to kill people." It's a weapon, and it's a weapon by virtue of the functions we care about (see U.S. Air Force, 2013). (Conversely, it's implausible to suggest that guns are "hunting implements that happen to be used in violence against humans"—the first 600 years of the history of firearms is of humans killing humans and can't be erased by its use in hunting game (LaFollette, 2018).)

Military robots occupy several roles and functions. Some of them, like many remotely piloted aircraft, are weapons. Others, however, serve functions like bomb detection, ordinance disposal, reconnaissance, or patrol. In the future, other functions are foreseen for robots. For example, the "Bloodhound" is a concept robot designed off the PackBot system developed by iRobot (as of 2016, Endeavor Robotics). The idea behind Bloodhound is to take some of the load off combat medics, whose job includes stabilizing and treating wounded warfighters on the battlefield (Singer, 2009). Bloodhound would be able to identify wounded soldiers, do basic diagnostics via video and attached devices such as a stethoscope, and administer some treatments like liquid bandages or painkillers.

These more sophisticated roles for robots, such as medic give us a reason to think of them more as members of *teams*. Where a weapon might be operated by a team—think of a large bomber, or a nuclear missile—that weapon isn't usually regarded as part of the team itself. Current robots are usually thought of as the same kind of tool—albeit important ones—such as the small remote-controlled car used by an Army sergeant to find explosive devices (ABC News, 2011). But as of 2009, the U.S. Department of Defense's intention has been, in the words of

Ethan Stump at the Army Research Laboratory, "trying to figure out what we can do to transition robots from tools to acting more as teammates within the squad (Ackerman, 2021)."

A QUESTION OF RELATIONS: WHAT ARE ROBOTS FOR—OR TO—US?

In considering shifts from tool to teammate, a change in how we *regard* an object raises an interesting question about how we should *treat* an object. After all, tools are disposable. Sure, I might maintain my tools and not want to lose them because of their utility. But we protect teammates. In fact, it is often wrong to allow our teammates to come to harm if we can prevent it from occurring.

This raises the question of the *defensive* obligations of robots and to robots. (Remember, I'm not claiming these robots have agency; this is at least for the near future a question of design.) Defense isn't easily distinguishable from offense, such as when we make weapons to defend ourselves that can also be used to attack others (Enemark, 2005). But we can distinguish the *obligation to defend* from the obligation to—or to not—attack.

Defense is often justified in one of two ways: Liability justifications, and lesser-evil justifications. Liability justifications hold that we are permitted to defend ourselves from another when they violate our rights not to be harmed (McMahan, 2002), or when they are responsible or (more restrictively) culpable for harm towards us. Lesser-evil justifications, on the other hand, hold that we are permitted to use force to defend ourselves from another when doing so would prevent an equal or greater harm (Frowe, 2014). Defending others has some wrinkles, particularly when it comes to our right to defend ourselves being transferrable to others, but the justifications are often seen to be the same.

It doesn't seem that we can use liability rights to justify defending robots—they at least for now don't have rights to be violated, and moreover can't be harmed in the way that humans and other sentient beings are harmed. But it seems intuitive to me that robots could be obligated to defend *us*, if they can. From my earlier caveat, this is a question of design, and it seems imminently plausible to me that we should program some robots to act in protection of our rights, especially if they can protect our right to life or right against being harmed. This would provide a reason for robots to engage in using force, even lethal force, in our defense. That, I'm sure, opens up concerns of the kind the Campaign to Stop Killer Robots is arguing against—allowing robots to autonomously engage in acts that might imperil humans.

Lesser-evil justifications, however, provide us with much stronger reasons to engage in defense of robots. Think back to Bloodhound, the medical robot. Presumably, shooting out that robot's legs would have the same consequence for humans in that conflict zone as wounding a combat medic in the same way. If the robot can't function, it can't help people who desperately need it. That would cost lives, and potentially a significant number of lives, from people who die without access to life-saving medical care. Defending the robot, or allowing it to defend itself when necessary, might prevent a lesser harm.

What counts as "necessary" does a lot of moral work here, however. When a robot is part of a team, for example, can it act to preserve that team's function, even if it means harming others? There are lots of easy cases we could imagine where it seems the answer is yes. Medical robots are an easy example, but we could also imagine the case of support robots in humanitarian missions. Consider robots, much like the PackBot on which the Bloodhound is based, that are carrying aid and necessary relief supplies to victims of a building collapse in a conflict zone—or maybe tools to continue digging people out of the rubble. In a case like this, where many lives are on the line and the robots are necessary to saving those lives, it doesn't seem difficult to imagine that it would not only be permissible but maybe even required for those robots to engage in lethal defense of their team and their team's mission, if they can. This would assign robots—and the team they part of—fairly standard rules of engagement: Fire only when fired upon, or in defense of civilian noncombatants (Lazar, 2016).

ROBOT WOLVES IN SHEEP'S CLOTHING: COMPLICATING THE APPLICATION

The worry, however, is that the dimensions of these obligations seem wildly open to abuse. Sure, we might say, a robot should defend a just cause. But the rules for defending and the those for establishing just cause in a particular mission—much less an entire war—surely aren't the same. To reduce complexity, let's imagine that we only care about what is legal under current interpretations of international humanitarian law. A robot would only need to be programmed to detect threats, and confirm them as legitimate targets, to act legally in defense of itself or its team. It may need to consult with military officials (e.g., the U.S. Judge Advocate General Corps) to confirm some ambiguities; but it might also be programmed to act in certain ways when those connections are interrupted or impossible. But—as complex as all that is, because we are not anywhere near that kind of robot yet—that is an order of magnitude less complex than giving robots a full set of international legal interpretations on particular kinds of missions and their effects on different kinds of populations, combined with national doctrine. And even that pales in comparison to trying to give a robot the political science, historical, and sociological knowledge to make robust decisions about the legality of war.

And all of that, of course, may end up with pacifist robots. In his *Contingent Pacifism: Revisiting Just War Theory*, Larry May notes that a 1992 International Criminal Court Case affirmed the human rights of enemy combatants and held that those rights are *not* waived upon entering a war (May, 2015). The consequence of this is that in cases where capture or nonlethal subdual is possible, we should *always* prefer this to lethal force. And—as proponents of autonomous weapons love to tell us (Arkin, 2008)—in principle, robots have more options than we do. Recent work with autonomous vehicles has shown it is possible for robot cars to detect risky behaviors in humans and either pace speeding cars or "catch" out-of-control vehicles, resulting in considerable reductions to the chance of death (Robinson et al., 2021). Why not in militaries, given the extraordinary funds that they receive for research and development, create potentially pacifist robots who engage in defensive but nonlethal actions that subdue or capture enemies, even at risk to themselves? After all, healing a robot's "injuries" is surely easier than it is for a human.

The reply, I suggest, is that the human creators and commanders of these robots do not *want* this kind of robot. That is, the point of the robot is that it will protect its team and do so with the narrowest calculus, acting "ethically" only in a sense that aligns with a very narrow account of what it means to act rightly. That's because military robots, according to scholars like Christian Enemark, are currently designed to make killing easier—not just easier to do, but easier to justify, far away from the eyes of the public (Enemark, 2013). In a team of humans, where credibility is on the line, it is hard to imagine deploying a robot that might consult international law and decide that it and its teammates ought not be permitted to use lethal defensive force, or that it ought to defend certain foreign civilians over its teammates.

The tension then lies in problems of practice and politics, compared to problems of principle. It seems clear that in some cases, in the right contexts, robots should be defended or defend themselves. They do not have defensive rights as moral agents, but defending a robot may nonetheless prevent some greater harm in virtue of that robot's function. In practice, however, and given the roles of drones in this world, however, it is very unclear whether that limited set of defensive opportunities justifies programming drones to exercise defensive prerogatives. This is because the fungibility between a good defense and a good offense, as they say, is immense.

MAKE THEM RARE OR MAKE THEM CARE

In a recent paper, my colleague and I mounted an argument against using AI in lethal, primarily *offensive* objectives based on moral cost sharing. We started with the idea that AI reduces the number of people who share in the moral cost of killing—that is, the weight of the act of killing. We then showed that increased moral cost often harms the human operators of military robots, and their controlling AI. This gives us a reason to avoid deploying AI that allow very small numbers of people to have control over large numbers of ways to kill other humans unless there was a suitable way to share the moral cost. One way this could be prevented, we concluded, is if we engineered AI to be able to share our moral costs—not full autonomy, but *sentience* (Hereth & Evans, 2023).

I want to extend that argument, based on the above, but away from moral cost sharing. Rather, we should take seriously the defensive prerogatives of robots, but in doing so should engineer robots with the necessary information to engage in acts of defense based on the full range of knowledge we have available to us, *even if that means they become pacifists.* It may be that robots turn out *true* pacifists and, with all the relevant defensive information on board, conclude that they should never engage in lethal defensive force. Or they may turn out *contingent* pacifists, wherein they will engage in defensive killing, just not in any conflict we have going right now (or, potentially, will in the conceive of in the future), because their analysis of the legality and justice of actual armed conflicts always leads them to avoid killing.

This sounds untenable. But why not? This is not a question posed to political scientists or sociologists: They will answer, correctly, that this is because it is unlikely that the institutions that fund this kind of robotic development want a robot that doesn't do what they want. This, rather, is a challenge to those institutions, especially those who talk about "value alignment" in artificial intelligences and robots. It is a provocation to put our moral money where our mouths are and train a robot to act in alignment with a comprehensive account of robust moral traditions around the world developed through hard lessons over hundreds or even thousands of years—rather than simply to align with the values of a narrow set of institutions today, which are subject to the vagaries of our ongoing political struggles. Why not embrace uncertainty of value alignment that leads to robots acting in ways we should but frequently fail to do: That the "crime of war" (Walzer, 2015) really is hell, and really is beyond the pale.

To return briefly to the specter of autonomy in robotic systems: Fifteen years of solid debate around "autonomous weapons" has neglected that, in philosophical writing around humans, autonomy requires some sense that the created have the capacity to make decisions beyond our creators. This need not be a religious sense of "created," either: Every human observes this, at least as child or as parent of other humans. Granting our robots the capacity to exercise such a morally serious capacity as defensive force requires giving them the opportunity to *refrain* from using it, even in cases where we might want them to do otherwise.

REFERENCES

ABC News. (2011, August 3). Afghanistan war: Hobbyists' toy truck saves 6 soldiers' lives. *ABC News*. https://abcnews.go.com/Technology/remote-controlled-truck-soldier-afghanistan-saves-soldiers-lives/story?id=14225434

Ackerman, E. (2021). How the U.S. Army is turning robots into team players. *IEEE Spectrum*. https://spectrum.ieee.org/ai-army-robots

Arkin, R. C. (2008). Governing lethal behavior: Embedding ethics in a hybrid deliberative/reactive robot architecture. In *Proceedings of the International Conference on Human-Robot Interaction* (pp. 121–128). ACM.

Campaign to Stop Killer Robots. (2023). *Stop Killer Robots*. https://www.stopkillerrobots.org/

Čapek, K. (2004). *R.U.R.* (C. Novack-Jones, Trans.). Penguin Classics. (Original work published 1920).

Coeckelbergh, M. (2010). Robot rights? Towards a social-relational justification of moral consideration. *Ethics and Information Technology*, *12*(3), 209–221.

Enemark, C. (2005). United States biodefense, international law, and the problem of intent. *Politics and the Life Sciences*, *24*(1–2), 32–42.

Enemark, C. (2013). *Armed drones and the ethics of war: Military virtue in a post-heroic age*. Routledge.

Frowe, H. (2014). *Defensive killing*. Oxford University Press.

Hereth, B., & Evans, N. (2023). Make them rare or make them care: Artificial intelligence and moral cost-sharing. In D. Schoeni, T. Vestner, & K. Govern (Eds.), *Ethical dilemmas in the global defense industry*. Oxford University Press.

LaFollette, H. (2018). *In defense of gun control*. Oxford University Press.

Lazar, S. (2016). *Sparing civilians*. Oxford University Press.

May, L. (2015). *Contingent pacifism: Revisiting Just War Theory*. Cambridge University Press.

McMahan, J. (2002). *The ethics of killing: Problems at the margins of life*. Oxford University Press.

Preston, B. (1998). Why is a wing like a spoon? A pluralist theory of function. *The Journal of Philosophy*, *95*(5), 215–54.

Robinson, P., Sun, L., Furey, H., Jenkins, R., Phillips, C. R. M., Powers, T. M., … & Evans, N. G. (2021). Modelling ethical algorithms in autonomous vehicles using crash data. *IEEE Transactions on Intelligent Transportation Systems*, *23*(7), 7775–7784.

Singer, P. W. (2009). *Wired for war: The future of military robots*. Brookings. Retrieved from https://www.brookings.edu/articles/wired-for-war-the-future-of-military-robots/

U.S. Air Force. (2013, April 4). MQ-1B PREDATOR [Fact Sheet]. https://web.archive.org/web/20130524052633/http://www.af.mil/information/factsheets/factsheet.asp?fsID=122

Walzer, M. (2015). *Just and unjust wars: A moral argument with historical illustrations*. Basic Books.

CHAPTER TWENTY SEVEN

Persons & Things: Rethinking the Ontology of the Robot

DAVID J. GUNKEL

Robots are a curious sort of thing. On the one hand, they are designed and manufactured technological artifacts. They are things. And like any of the other things that we encounter and use each and every day, they are objects with instrumental value. Yet, and on the other hand, these things are not quite like other things. They have social presence, they are able to talk and interact with us, and many are designed to mimic or simulate the capabilities and behaviors that are commonly associated with human or animal intelligence. Robots therefore invite and encourage zoomorphism, anthropomorphism, and even personification.

So, are robots things—technological objects that we can use or even abuse as we decide and see fit? Or is it the case that robots can or even should be something like a person, i.e., another subject who would need to be recognized as a kind of socially significant Other who would require our respect? These questions, which have been a staple in science fiction since the moment the robot stepped foot on the stage of history—quite literally in this case, since the word "robot" is initially the product of a 1920 stage play by Czech playwright Karel Čapek—are no longer a matter of fictional speculation. These questions are now a very real legal and philosophical problem (Gunkel, 2022a).

Resolving this seems pretty simple. All that would be needed is to assemble the evidence, develop a convincing case, and then decide whether to categorize robots as one or the other. This is not just good reasoning, it's the law. In fact, the binary distinction separating who is a person from what is a thing has been the ruling conceptual opposition in both moral philosophy and jurisprudence for close to 2000 years. When the Roman jurist Gaius (130–180 CE), in a treatise he titled *Institutes*, explained that law involved two kinds of entities, either persons or things, he instituted a fundamental ontological division that has been definitive of Western (but not just Western) moral and legal systems.

In the face of others—another human being, a nonhuman animal, a tree, an extraterrestrial, a robot, etc.—the first and perhaps most important question that must be addressed and resolved is "What is it?" Is it another *subject* similar to myself, to whom I would be obligated? Or is it just an *object* that can be taken-up, possessed, and used without any further consideration or concern? Consequently, all that is needed is to decide whether robots are things or persons. Sounds easy enough. But this determination turns out to be much easier said than done. In fact, the robot does not quite fit or easily accommodate itself to either category (Gunkel, 2022a).

THINGS

Classifying robots as "things" seems to be the easiest of matters. We all kind of know what things are, such that asking a question like "What is a thing?" seems to be impertinent and immaterial. But as the German philosopher Martin Heidegger pointed out, this is precisely the problem. Because things are already familiar—perhaps too familiar—we have little or no critical distance on them as things. This is because, as Heidegger (1927/1962) explains, things are not typically disclosed to us as things but encountered as objects. In other words, things are not experienced as mere entities laying around out there in the world. They are always pragmatically situated and characterized in terms of our involvements and interactions with the world in which we live. For this reason, things are first and foremost made available to us or revealed as *objects*. They are objectified.

Robots, however, do not seem to fit easily within this framework. Consider, for example, what is now a rather common but still surprising social practice. Users of digital voice assistants, like Siri and Alexa, often find themselves saying "thank you" to the object. This is both weird and disorienting. We typically do not express gratitude to things or feel bad about not doing so. We use automobiles to travel around town without feeling the need to say "thank you" to the vehicle. But if we take a taxi or use a ride sharing service, we will—or we think we should—say "thank you" to the driver of the vehicle, who we recognize as another person. Because digital voice assistants are things that talk like another person, we often (and rather non-consciously) respond to the object as if it were something other than a mere thing, a kind of someone to whom we feel obliged to say "thank you" (Gunkel, 2022b, p. 734).

PERSONS

Asking "can robots be persons?" again seems to be direct and immediately understandable. But as was the case with things, the concept of "person" also turns out to be far more interesting and complicated than one might anticipate, making answers to this inquiry equally difficult to resolve.

Typically, when we use the word "person," we are referring to another human being. But this seemingly natural and everyday understanding is not entirely accurate. The word "person" has a long and intricate history. It is originally derived from the Latin noun *persona*, which referred to the mask worn by actors in a stage play. As a result of this, "person" denotes not just an individual human being but more accurately names the role that one plays or is assigned within the context of a social situation or performance. This can be seen especially in law, where who is considered a person has not been limited to human individuals but can also be extended to other kinds of non-human entities that occupy or play roles that are important for our shared social reality—corporations, organizations, ships, animals, and the natural environment. In order to attend to this important difference, person is often divided into two types: *natural person* and *legal person*.

Natural Person

Whether something is or is not a natural person is usually decided on the basis of individual properties and capabilities—e.g., rationality, consciousness, sentience, etc. So there is a kind of litmus test for achieving recognition as a person. We (and who is interpellated by this first-person-plural pronoun will not be insignificant) first define the criteria for what makes an entity a person. In other words, we devise a standard by deciding what we believe are the necessary person-making properties or capabilities. We then use this standard to test and evaluate whether some entity qualifies as a natural person or not. So the question—whether robots or AI could ever become natural persons—is a question that we get to evaluate based on criteria that we get to define.

Responses to this question typically divide into a simple and mutually exclusive yes/no set of rejoinders. On the side of "no," there have been efforts to protect human exceptionalism by limiting who can be included in the category person. According to this way of thinking, no matter how sophisticated robots and AI may become or seem to be, they will never actually achieve any of the person-making properties, like consciousness, rationality, or sentience. They might appear to do so through various behaviors that we take as signs of an inner life (e.g., consciousness or sentience), but they do not and cannot really possess the capability that they have been designed to simulate. On the other side, you have those who assert the exact opposite, arguing that, "yes," artifacts like AI, robots, and autonomous systems, either are or will be able to achieve the benchmarks of person. If and when—and it is more often than not a matter of "when"—this happens, then withholding the title of person from these things would be unethical and unjust.

Perhaps the best example of this tension and its significance can be seen in the events surrounding former Google engineer Blake Lamoine and the LaMDA large language model. In June of 2022, Lemoine claimed that the LaMDA system was conscious and therefore was a person deserving of moral respect and legal consideration. Google—the corporation that funded the research and built LaMDA—took the opposite position, arguing that LaMDA, like any computer application, was not conscious and then suspended and eventually fired Lemoine for his claims. Both sides in this debate asserted and sought to justify their positions by mobilizing what Mark Coeckelbergh (2012, 2014) has called the "properties approach" to deciding moral status. And each side struggled with the same set of problems—accurately defining the qualifying ontological or psychological properties and then proving (or disproving) that these properties were actually present in the algorithm. In fact, the entire debate was defined by and circulated around the inability to resolve these issues (Gunkel, 2022c).

Legal Person

Unlike a natural person, a status grounded in the ontological conditions or essential properties of the individual entity, a legal person is a socially constructed and conferred recognition. In other words, something becomes someone not because of their essential nature but due to the fact that they have been recognized by others as having a particular status or social role. To be a person, then, means that one comes to be recognized as a subject under the law, possessing both responsibilities and rights within a particular legal construct or institution. If the paradigmatic natural person is the human being in possession of a set of natural capabilities that make one a person, the paradigmatic legal person is the corporation, which is a person not by its intrinsic nature but due to the fact that it is recognized and situated within the law as a subject of the law.

Thus, the question confronted in the face (or the faceplate) of the robot or other seemingly intelligent and/or socially interactive artifact is whether it would make sense to extend the category of legal person to these other kinds of entities. Responses to this question also tend to divide into two opposed and seemingly irreconcilable positions. Those who are critical of the idea argue that extending the recognition of person to these technological artifacts, although clearly possible and entirely legal, is wrong and should not be allowed to happen. Those who have advocated for it argue that extending legal personhood or personality to robots and other intelligent (or at least seemingly intelligent) artifacts will be necessary for integrating these technologies into our social reality and moral/legal systems.

The terms of this debate became evident in May of 2016, when the Committee on Legal Affairs of the European Parliament—the legislative branch of the European Union—proposed that "sophisticated autonomous robots" be considered "electronic persons" with specific rights and obligations for the purposes of contending with the challenges of technological unemployment, tax policy, and legal liability (Delvaux, 2016; Gunkel, 2022b). The proposal, which did not pass as originally formulated, was strongly opposed by over 250 experts in the field of AI and robotics, who signed onto an open letter addressed to EU lawmakers. According to the opinion expressed in this document, the very idea of "creating a legal personality for a robot is inappropriate whatever the legal status model" (Robotics Open Letter, 2017).

A MORAL AND LEGAL ONTOLOGY FOR THE 21ST CENTURY

The person/thing dichotomy has been an undeniably useful and influential ordering principle, one that not only has the weight of history behind it but has been codified in language and law. For this reason, the principal challenge that is now being confronted in the face of robots and AI concerns how *we* decide to fit these artifacts into this often unquestioned and seemingly unassailable ontological order. The problem is that accommodating these technological innovations to one category or the other has so far been inconclusive and seemingly irresolvable. "In the dichotomous model that has long opposed the world of things to the world of persons," Esposito (2015, p. 3) writes "a crack appears to be showing." And "robot" designates that fissure. So now what? How can or should we respond to this challenge or opportunity? Let me conclude by briefly considering three possible modes of response.

Status Quo

We can decide to reassert and defend the existing moral and legal ontology, forcing robots and AI to fit into one or the other designation. As Esposito (2015, pp. 1–2) has explained: "If there is one assumption that seems to have organized human experience from its very beginnings it is that of a division between persons and things… Since Roman times, this distinction has been reproduced in all modern codifications, becoming the presupposition that serves as the implicit ground for all other types of thought—for legal but also philosophical, economic, political, and ethical reasoning." Proceeding according to this standard way of organizing things (actually not just things but things and persons) seems reasonable and natural, but that is the problem and for two reasons.

First, in this binary contest of either/or, both sides advance positions and arguments that—at the time one initially hears them—seems to make good sense. And like similar polarizing disagreements—think, for example, of other irresolvable moral or legal disputes, like the abortion debate or physician-assisted suicide—there has been no clear winner. Both sides continue to heap arguments and evidence in support of their position, but the basic terms and conditions of the conflict remain largely in place and essentially unchanged (Gunkel, 2022b). As such, the debate has lead thinking and decision making into a kind of *cul-de-sac* or what German philosopher G.W.F. Hegel (1816/2010, p. 202) called "a bad infinity."

Second, the person/thing dichotomy is not natural. It is an artifact produced by and belonging to a particular culture and its specific moral and legal philosophy. Other cultures and traditions, distributed across time and space, do not divide up and make sense of the diversity of beings in this arguably binary fashion. They perform decisive cuts separating the *who* from the *what* according to other ways of seeing, valuing, and acting. As Archer Pechawis explains in his contribution to the essay "Making Kin with Machines": "*nēhiyawēwin* (the Plains Cree language) divides everything into two primary categories: animate and inanimate. One is not 'better' than the other, they are merely different states of being. These categories are flexible: certain toys are inanimate until a child is playing with them, during which time they are animate. A record player is considered animate while a record, radio, or television set is inanimate. But animate or inanimate, all things have a place in our circle of kinship or *wahkohtowin*" (in Lewis et al., 2018). This alternative formulation—which is just one among many—runs counter to the dominant ways of proceeding, seeing the boundary between what Western philosophy and law calls "person" and "thing" as being endlessly flexible, permeable, and more of a continuum than an exclusive, binary opposition.

Third Alternatives

One frustration with conceptual oppositions, like that which has distinguished person from thing, is that they have "compressed and continue to compress human experience into the confines of this exclusionary binary equation" (Esposito, 2015, p. 4), such that between person and thing "there appears to be nothing" (Esposito, 2015, p. 16). Though binary oppositions have a certain functionality and logical attraction, they often seem to be unable to represent or to capture the rich experiences of actual existing empirical reality, which always seems to complicate simple reduction into one of two options. It is for this reason that we are generally critical of "false dichotomies"—the parsing of complex experience into a simple and irreducible either/or distinction.

One method for resolving this problem is to formulate a third alternative that is neither one thing nor the other but a kind of combination or synthesis of the one and the other. Alternatives like this sound liberating and hold considerable promise, precisely because they appear to interrupt the structural limitations imposed by either/or logic and arrange for a more nuanced representation and understanding that is both/and. Such an alternative is already available to us with the legal institution of slavery, and there are a number of researchers who have argued that "robots should be slaves" (Bryson, 2010; Pagallo, 2011).

Already in Roman times, slaves were regarded as something more than a mere thing but not quite a full person. They occupied a position that was situated in between the one and the other, being both thing and person (Esposito, 2015). And there has been, in both the legal and philosophical literature, a surprising number of serious proposals arguing for instituting what can only be called Slavery 2.0 (Gunkel, 2018). Clearly use of the term "slave" is provocative and morally charged. It would be impetuous to presume that the various proposals for repurposing the paradigm of slavery to deal with robots and AI would be the same or even substantially similar to what had occurred (and is still unfortunately occurring) with human bondage. But, and by the same token, we also should not dismiss or fail to take into account the documented evidence and historical data concerning slave-owning societies and how institutionalized forms of slavery affected both individuals and human communities. The corrupting influence of socially sanctioned, institutionalized bondage concerns not just the enslaved population but also those who would occupy the position of mastery.

Deconstructing Things

There is no doubt that robots and AI are technological things, but they are not necessarily the kind of things that are (or should be) situated and conceptualized as the polar opposite of persons. They are not limited to being merely objects for a subject, instrumental means as opposed to an end, or thing in distinction to person. Instead, it is with these things that "the bivalent logic of modernity is opening up to other paradigms"—other ways of thinking and responding in the face of "things that are no longer merely objects, and of subjects who are increasingly difficult to confine inside the dispositif of the person" (Esposito, 2015, p. 132).

Consequently, robots and AI are a kind of glitch or anomaly that interrupts and deconstructs the existing binary logic that differentiates person from thing. They destabilize the very terms and conditions of the debate, and in doing so they invite us to rethink the existing categories. Of note who is and what is not interpellated by this first-person plural "us" is itself part of the problem. It is, therefore, in the face of the robot that we are called to respond to and take responsibility for beings that are other—that do not fit or accommodate the existing moral and legal ontology.

But that means that all of this is not really about robots, AI, and other artifacts. It is about us. It is about the moral and legal structures we have fabricated to make sense of and order all things. And it is with the robot that we are now called to take responsibility for this privileged situation and circumstance. Thus, what is needed in response to this opportunity or challenge is not some forceful reassertion of the usual ways of thinking but a significantly reformulated moral and legal ontology that can scale to the opportunities and the challenges of the 21st century and beyond.

What this new ontology will look like and how it will be developed is the task for thinking and acting from this point forward. Confronting this will be as terrifying and exhilarating as any of the robot uprisings that have been imagined in science fiction, because getting this right will require nothing less than a thorough rethinking of everything we thought was right, natural, and beyond question. But it matters. It matters for us, and it matters for the other entities—both naturally occurring and artificially made—that reside on and share this fragile planet with us.

REFERENCES

Bryson, J. J. (2010). Robots should be slaves. In Y. Wilks (Ed.), *Close engagements with artificial companions: Key social, psychological, ethical and design issues* (pp. 63–74). John Benjamins.

Coeckelbergh, M. (2012). *Growing moral relations: Critique of moral status ascription*. Palgrave MacMillan.

Coeckelbergh, M. (2014). The moral standing of machines: Towards a relational and non-Cartesian moral hermeneutics. *Philosophy and Technology, 27*, 61–77.

Delvaux, M. (2016). Draft report, with recommendations to the Commission on Civil Law Rules on Robotics, 2015/2103(INL). Committee on Legal Affairs. European Parliament. https://www.europarl.europa.eu/doceo/document/JURI-PR-582443_EN.pdf

Esposito, R. (2015). *Persons and things* (Z. Hanafi, Trans.). Polity.

Gunkel, D. J. (2018). *Robot rights*. MIT Press.

Gunkel, D. J. (2022a). In the face of the robot. *Communication +1, 9*(1). https://doi.org/10.7275/8bqq-2h34

Gunkel, D. J. (2022b). Should robots have standing? From robot rights to robot rites. In R. Hakli et al. (Eds.), *Social robots in social institutions* (pp. 733–738). IOS Press.

Gunkel, D. J. (2022c). The relational turn: A media ethics for the 21st century and beyond. *Media Ethics Magazine, 34*(1). https://www.mediaethicsmagazine.com/index.php/browse-back-issues/219-fall-2022-vol-34-no-1/3999399-the-relational-turn-a-media-ethics-for-the-21st-century-and-beyond

Gunkel, D. J. (2023). *Person, thing, robot: A moral and legal ontology for the 21st century and beyond.* MIT Press.

Hegel, G. W. F. (2010). *The science of logic* (G. Di Giovanni, Trans.). Cambridge University Press. (Original work published 1816)

Heidegger, M. (1962). *Being and time.* (J. Macquarrie & E. Robinson, Trans.). Harper & Row. (Original work published 1927)

Lewis, J. E., Arista, N., Pechawis, A., & Kite, S. (2018). Making kin with the machines. *Journal of Design and Science, 3.5.* https://doi.org/10.21428/bfafd97b

Pagallo, U. (2011). Killers, fridges, and slaves: A legal journey in robotics. *AI & Society, 26,* 347–354.

Robotics Open Letter. (2017). Open letter to the European Commission Artificial Intelligence and Robotics. http://www.robotics-openletter.eu/

This chapter is based on and derived from research that was originally presented in Gunkel (2023).

Index of Social Robots

Aibo	(Sony Corporation, Japan) refers to one of varied models of robotic dogs first released in 1999. https://us.aibo.com/
Ameca	(Engineered Arts, U. K.) is a genderless humanoid robot with rubber skin, first released in 2021. https://www.engineeredarts.co.uk/robot/ameca/
ANYmal	(ANYbotics, Switzerland) is a four-legged, industrial-grade robot marketed primarily for its ability to traverse varied terrain, sense environments, and carry payloads. The is also an "X" version that is marketed as safe in explosive atmospheres. https://www.anybotics.com/robotics/anymal/
ASIMO	(Honda, Japan) was a white-colored humanoid robot standing just over 50 inches tall. The robot was made public in 2000; production ceased in 2018. https://global.honda/en/robotics/asimo/
ATRIAS	(Oregon State University Dynamic Robotics Laboratory, U.S.A.) is short for "Assume the Robot is a Sphere." It is a bipedal robot modeled after humans and animals that have two-legged locomotion. https://films.oregonstate.edu/atrias
BigDog	(Boston Dynamics, USA) is a quadrupedal machine primarily developed for military use but was ultimately never deployed. It was Boston Dynamics first robot to operate outside its laboratory. https://bostondynamics.com/legacy/

Bloodhound	(iRobot, formerly Endeavor Robotics, now a division of Teledyne Flir, USA) was a robotic medical assistant whose development was funded by the Department of Defense. It was designed to "navigate autonomously to wounded soldiers and provide telepresence for remote medics." It followed the tank-like template of iRobot's PackBot (see entry below). https://www.sbir.gov/content/robotic-medic-assistant-0
Cassie	(Oregon State University Dynamic Robotics Laboratory, USA) was a bipedal robot modeled after the cassowary (a flightless bird), with backward-directed knees. It set the record in 2022 for the quickest traverse of 100 meters by a robot—24.73 seconds. https://www.roboticsproceedings.org/rss14/p54.pdf
Cozmo	(Anki, Digital Dream Labs, USA) is a small unit with a core on treaded wheels, an animated face, and dump-truck like appendage that moved up and down; it was accompanied by lighted blocks that it could interact with. Production ceased in 2019 due to bankruptcy but relaunched when Digital Dream Labs acquired the intellectual property. https://ddlbots.com/products/cozmo-robot
Mavic	(DJI, China) is a remote-controlled aerial drone with a high-resolution camera; its X-shaped frame features a central hub with power and computing equipment, while each of the four arms feature a motor and propeller. https://www.dji.com/products/camera-drones#mavic-series
Da Vinci Surgical Robots	(Intuitive, USA) are a class of machines designed to augment human surgeons' performance and precision during surgical procedures. More recent versions have incorporated machine vision and data analytics functionality. https://www.intuitive.com/
Dragonbot	(MIT Personal Robotics Group, USA) is a dragon-shaped base unit that uses a mobile phone to capture sensory input and to control its behaviors. https://www.media.mit.edu/projects/dragonbot-android-phone-robots-for-long-term-hri
EMO	(Living AI, China) is a desktop robot pet with more than 1,000 digital facial expressions and bodily movements. It consists of a cube-like body, two legs, and a pair of headphones. https://living.ai/emo/

ERICA	(Osaka University, Kyoto University, Japan) is a gynoid that can understand and synthesize natural language, detect specific human behaviors, and decide on appropriate behaviors given context. https://www.youtube.com/channel/UCDjRgo5ecEw0Ou78-uJOssg
Flobi	(Bielefeld University, Germany) is a cartoonish robotic head, with interchangeable features. It has a moveable mouth and eyebrows to suggest different expressions and is designed for interactions with young people. http://dx.doi.org/10.1109/ROBOT.2010.5509173
Harmony	(RealDoll, USA) is a customizable sex robot that can form facial expressions, speak, and through accompanying hardware can be ascribed an individual personality. https://www.realdoll.com/product/harmony-x/
Huggable	(MIT Personal Robotics Group, USA) is a plush, teddy bear-like robot characterized as a robotic companion to participate in triadic interactions—primarily in healthcare and educational environments. https://www.media.mit.edu/projects/huggable-a-social-robot-for-pediatric-care/overview/
Jibo	(MIT Personal Robotics Group, USA) was marketed as the first social robot for the home and as a research platform. It featured a simple torso and head, with a screen-presented face consisting only of a round, animated eye that sometimes shifted into simple icons. Its creator company stopped support for Jibo in 2019 and the robot famously announced its own demise. https://techcrunch.com/2019/03/04/the-lonely-death-of-jibo-the-social-robot
K5	(Knightscope, USA) is an autonomous, amorphously shaped robot designed to autonomously support property security efforts—detecting threats and anomalies, delivering alerts, and serving as a deterring presence. https://www.knightscope.com/products/k5
Kargu-2	(STM, Turkey) is an aerial combat drone that relies on human control, capable of identifying and striking targets that are stationary or moving. https://www.stm.com.tr/en/kargu-autonomous-tactical-multi-rotor-attack-uav
Kaspar	(University of Hertfordshire, United Kingdom) is a child-sized humanoid robot designed to support the lives of children with

286 | INDEX OF SOCIAL ROBOTS

	autism or other communicative challenges. Its soft, gender-neutral face uses simplified expressions and features. https://www.herts.ac.uk/kaspar/the-social-robot
Keepon	(BeatBots, USA) is a minimalist robot with a morphology similar to a snowman—two stacked, yellow spheres sitting atop a black base; it has two simple eyes and a black microphone nose. It is primarily developed for use as a therapeutic device, especially for children with developmental disorders. https://beatbots.net/keepon-pro
KeJia	(University of Science and Technology of China) is a gynoid that communicates primarily by voice through speech recognition and synthesis and moves through environments with a wheeled base. http://www.doi.org/10.1007/978-3-319-25554-5_15
Kinova Gen3	(Kinova, Canada) is a lightweight robotic arm with 6 or 7 degrees of freedom, depending on the model. https://www.kinovarobotics.com/product/gen3-robots
Kirobo Mini	(Toyota, Japan) is a mini robot companion—about 10 cm tall—that can have conversations with gestures. It can fit in its soft carrying case or in a vehicle cupholder. https://pressroom.toyota.com/toyota-kirobo-mini-sales/
Kismet	(MIT Humanoid Robotics Group, USA) is a robotic head whose design emphasizes communicative features: Blinking eyes and mobile eyebrows, moving mouth, and conspicuous ears anchored to a metal frame. http://www.ai.mit.edu/projects/humanoid-robotics-group/kismet/kismet.html
Kuratas	(Suidobashi Heavy Industry, Japan) is a humanoid mech (or mecha—a human-piloted robot common to anime and manga). The user sits inside the machine and controls it with a Microsoft Kinect-based device, driving it forward and backward on the robot's four wheels. https://www.reuters.com/article/idUSBRE8AS04G/
Leka	(Leka, France) is a spherical robot, about 18 cm across, with an animated screen fact that displays simplified facial expressions. The robot is designed to facilitate childhood cognitive and affective learning and play, especially for children requiring personalized interactions. https://leka.io/en/who-is-leka/

INDEX OF SOCIAL ROBOTS | 287

Mini-Cheetah (MIT Biomimetic Robotics Lab, USA) is a quadrupedal robot unique for its agility, able to do backflips and walk in multidirectional patterns.
https://news.mit.edu/2019/mit-mini-cheetah-first-four-legged-robot-to-backflip-0304

Moxi (Diligent Robotics, USA) is a demi-humanoid robot with a torso, a single arm and hand, a head, and a cylindrical base. It was designed to facilitate healthcare work primarily by delivering supplies; it became more broadly known for delivering PPE during the COVID-19 pandemic.
https://www.diligentrobots.com/moxi

Moxie (Embodied, Inc., USA) is a small, humanoid robot with an expressive animated face designed to facilitate academic learning and practice communicating and mindfulness.
https://moxierobot.com/

MQ-1 Predator (General Atomics Aeronautical Systems, USA) is one of a series of remote-piloted aerial drones. The original RQ-1 was designed for surveillance, and its designation was changed to MQ-1 with the M to reflect "multiple" use cases when it became an armed aircraft.
https://www.airforce-technology.com/projects/predator-uav/

NAO (Aldebaran Robotics, previously Aldebaran Robotics, Japan/France) is a small humanoid robot that can speak, walk, sense environment elements, and recognize objects. It is the robot famously used in the RoboCup tournament in which robots play soccer.
https://us.softbankrobotics.com/nao

PackBot (Teledyne Flir, USA) is a field robot marketed for bomb disposal, surveillance, and hazmat operations. It is shaped somewhat like a tank, with a tracked locomotion, a manipulator that can lift, sense, or perform other functions through specific accessories.
https://www.flir.com/products/packbot-525/

PaPeRo (NEC Corporation, Japan) is a small robot with a monolithic body and movable head. Its name stands for Partner-type Personal Robot, it can recognize faces through a pair of cameras and will attempt to start a conversation when a face is detected.
https://www.necplatforms.co.jp/solution/papero_i/

Paro (AIST, Japan) is shaped like a seal, with soft white fur on its exterior. It learns from user behavior to adapt its own actions to user preferences and has been designed as a therapeutic companion for older adults.

288 | INDEX OF SOCIAL ROBOTS

Pepper
    http://www.parorobots.com/
    (SoftBank Robotics, previously Aldebaran Robotics, Japan/France) is a tall, white-shelled humanoid robot featuring a screen on its chest and wheel-based locomotion.
    https://us.softbankrobotics.com/pepper

Perseverance
    (NASA, USA) is one in a series of rovers built to "seek signs of ancient life and collect samples of rock and regolith" as part of scientific explorations of other planets.
    https://mars.nasa.gov/mars2020/

Pleurobot
    (Swiss Federal Institute of Technology, Switzerland) is a robot inspired by the salamander, *Pleurodeles waltl*, for both its foot design and limb motion.
    https://doi.org/10.1098/rsif.2015.1089

RABBIT
    (The Laboratory of Digital Sciences of Nantes, France; with Biped Robotics Lab, University of Michigan, USA) is a robotic machine that simulates human walking (sans feet) and was created to better understand bipedal locomotion.
    https://www.biped.solutions/rabbit

Raibo
    (Railab, Korea) is a quadrupedal robot that can dynamically maneuver over varied terrain, including hills and stairs.
    https://www.railab.kaist.ac.kr/sections/robots

Reachy
    (Pollen Robotics, France) has a humanoid upper body and a minimalist lower body consisting of a wheeled base and a pole on which the upper body is mounted. It is remotely operated by VR.
    https://www.pollen-robotics.com/reachy/

ReBeL
    (igus, Germany) stands for Robotic embedded-BDLC & electronics Link. It is a collaborative robot with an arm-like morphology.
    https://www.igus.com/info/rebel-cobot

RHex
    (orig. University of Michigan, USA; McGill University, Canada;) robots are hexapods (six-legged units) designed with a level, flat surface on which instrumentation can be mounted; in particular it is used for study of wind phenomena.
    https://www.rhex.web.tr/

RoboThespian
    (Engineered Arts, U.K.) is an adult human-sized android originally designed to give stage performances. It features a white outer shell, articulated joints from the waist up, and animated eyes; it has limited autonomous behavior or can be controlled remotely or pre-programmed.
    https://www.engineeredarts.co.uk/robot/robothespian/

INDEX OF SOCIAL ROBOTS | 289

Roomba	(iRobot, USA) is an automated vacuum cleaner, with object avoidance and scheduling, with more recent models featuring automated dirt disposal, mopping functions, https://www.irobot.com/roomba
Sophia	(Hanson Robotics, Hong Kong) is an adult-sized gynoid run by a combination of AI technologies that creators claim simulates human emotions and may have a rudimentary form of consciousness. She is recognized as the first robot Innovation Ambassador for the United Nations Development Programme. https://www.hansonrobotics.com/sophia/
Spot	(Boston Dynamics, USA) is a dog-inspired quadruped with the ability to traverse varied terrain, carry payloads, and sense and monitor environments when equipped with optional instrumentation. https://bostondynamics.com/products/spot/
StarlETH	(Robotics Systems Lab, ETH Zürich, Switzerland) was a quadrupedal robot with four symmetric legs each with three degrees of freedom to mimic hip and knee movements. https://rsl.ethz.ch/robots-media/starleth.html
Tay	(Microsoft, USA) was a chatbot deployed on Twitter in 2016 that learned from the language it encountered on the platform. It famously produced racist, sexist, and antisemitic after ostensible internet trolls fed it antisocial content. https://archive.org/details/TayTweets_201712
Turtlebot	(Melonee Wise and Tully Foote, USA) is an open-source robot kit with an iRobot Create base, featuring location mapping and navigation. It was originally a cylindrical unit, but more recent versions have customizable shapes. https://www.turtlebot.com/about/
WABOT	(Waseda University, Japan) was a humanoid robot design, designed in 1970 and claimed as one of the first functioning anthropomorphic robots. It had the basic shape and functionality of a human body, able to verbally communicate, measure distance and direction, walk, and grip objects. https://www.humanoid.waseda.ac.jp/booklet/kato_2.html
WL-10RD	(Waseda University, Japan) was a walking robot developed in 1984 and is claimed as the first robot to dynamically and completely walk, at 1.3 seconds per step. https://www.humanoid.waseda.ac.jp/booklet/kato_4.html

Notes on Contributors

Sara Ali (Ph.D., National University of Sciences and Technology) is Associate Professor at the School of Mechanical and Manufacturing Engineering, Department of Robotics and AI at National University of Sciences and Technology (Pakistan). Her research focuses on human-robot interaction and trust, with an emphasis on social robots, AI, VR, and human-in-loop systems. She is the director of the joint international lab NUST-COVENTRY Human-Robot Interaction and the scientific director of national lab IFRL under the National Center of Artificial Intelligence. Sara's favorite robot is TARS—a standout character from *Interstellar* which is a loyal and assistive robot companion in their mission.

Jason Edward Archer (Ph.D., University of Illinois at Chicago) is Assistant Professor of Communication and Media Technologies at Michigan Technological University (USA). His research is situated at the nexus of communication, media studies, science and technology studies (STS), and sensory studies, with emphasis on human-machine communication and haptics. He directs the Human Machine Culture research group (Institute of Policy, Ethics, and Culture) and he co-edited the special issue Haptic Media in *New Media and Society*. Jason currently spends time in the office chatting with LivingAI's EMO robot about his next big project.

Jaime Banks (Ph.D., Colorado State University) is Associate Professor and Katchmar-Wilhelm Endowed Professor at the School of Information Studies at

Syracuse University (New York, USA). Her research focuses on the dynamics and effects of human-machine relations, with an emphasis on social robots, artificial intelligence, and videogame avatars. She is the director of the iSchool's LinkLab, and her work on social cognition and trust in human-robot interaction has been funded by the U.S. Air Force Office of Scientific Research. Jaime's favorite social robot is only *kind of* a robot—the animated suit of armor known as "Alphonse" from the *Fullmetal Alchemist* manga and anime series.

Nick Bowman (Ph.D., Michigan State University) is Associate Professor in the S.I. Newhouse School of Public Communications at Syracuse University (New York, USA). His research considers the cognitive, emotional, physical, and social demands of interactive media, with a focus on video games and extended reality technologies. He has published more than 150 peer-reviewed manuscripts, and has faculty affiliations in Taiwan, Mexico, and Canada. His favorite robot is BMO, the sentient video game from *Adventure Time*.

Andreas Butz (Dr., Saarland University) is Professor at LMU Munich (Germany) and leads the research group for Human-Computer Interaction at that institution. His research looks at interaction beyond the desktop, for example with robots, cars, or entire living spaces. His work on human-centered intelligent environments received the Alcatel-Lucent research award on technical communication in 2007. From the robots in popular culture, he likes C3PO best, for its civilized manners, surprising language skills, and humoristic talents.

Antonio Chella (Ph.D., University of Palermo) is a full Professor at the Department of Engineering at the University of Palermo (Italy). His research focuses on robot consciousness and cognition, with an emphasis on social robots and human-robot interactions. He is the director of the RoboticsLab, and his work on robot inner speech in trustworthy human-robot interactions has been funded by the U.S. Air Force Office of Scientific Research. Antonio's favorite robot is Andrew Martin (NDR-114), the main character of the movie *Bicentennial Man*, starring Robin Williams and based on Isaac Asimov's novels.

Chien-Hsiung Chen (Ph.D., University of Kansas) is Professor in the Department of Design at the National Taiwan University of Science and Technology (Taiwan). His cross-disciplinary research interests include industrial and interaction design, cognitive design, design research methodology, and user experience research and design. Dr. Chen received numerous research grants from the National Science and Technology Council (NSTC) for his studies. He used to be the Dean of the College of Design at Taiwan Tech and is currently the President of the Chinese

Institute of Design. Dr. Chen's favorite robots are C-3PO and R2-D2 from the *Star Wars* movie series.

Chris Chesher (Ph.D., Macquarie University) is Senior Lecturer in Digital Cultures in the Discipline of Media and Communications at the University of Sydney (Australia). His research explores digital media and robots through an interdisciplinary lens, including visual culture, assemblage theory, media studies and science and technology studies. His book *Invocational Media: Reconceptualising the Computer* is published by Bloomsbury. Chris's current favorite robot is Bellabot, a cute and hard-working restaurant service robot with an animated cat face that brings food to customers' tables.

Sarah Diefenbach (Dr., University of Koblenz-Landau) is professor for Economic Psychology and Human-Computer Interaction at the Department of Psychology at LMU Munich (Germany). Her research group explores design factors and relevant psychological mechanisms in the context of interactive technology and digital society, e.g., social robots, social media, digital collaboration. Sarah's favorite social robot is 'MM7 from the *Columbo* movie *Mind Over Mayhem* when computer science was obviously still so simple that a (smart) child could construe a robot.

Dariusz Doliński (Ph.D., Technical University of Darmstadt) currently serves as Professor and Head of the Department of Social Psychology at SWPS University in Wrocław (Poland). He is also a distinguished member of the Polish Academy of Sciences. His research is dedicated to exploring various aspects of social influence, with a particular focus on obedience to authority, the phenomenon of mimicry, and techniques of social influence. Dariusz's favorite social robot is C3PO, a beloved humanoid robot character from the *Star Wars* franchise, who makes appearances in each installment of the original trilogy.

Autumn Edwards (Ph.D., Ohio University) is Professor of Communication at Western Michigan University (USA), specializing in human-machine communication and human-robot interaction with a focus on ontology and communication metaphysics. She co-directs the Communication & Social Robotics Labs (combotlabs.org) and is the founding Editor-in-Chief of the journal *Human-Machine Communication* (hmcjournal.com). Autumn's favorite robot is CHAPPiE from the eponymous movie, for embodying the blurred boundaries of human and machine beings and highlighting their co-evolution in a shared ecology.

Nicholas G. Evans (Ph.D., Australian National University) is Associate Professor of Philosophy at the University of Massachusetts Lowell (USA). He works at the intersection of the ethics of emerging technology, national security, and public health, with a focus on the way that advances in the life and computer sciences impact how we govern international health and security. His current work focuses on the ethics of human enhancement and is funded by the US Air Force Office of Scientific Research, and the Greenwall Foundation. His favorite robot is Evangelion Unit 1; while he admits that bioengineered humanoids are only loosely a "robot," he's willing to die on this hill because he thinks that's where we'll ultimately take technological convergence.

Leopoldina Fortunati is Senior Professor at the Department of Mathematics, Computer Science, and Physics at the University of Udine (Italy). Her research primarily focuses on the cultural and social aspects of the diffusion of social robots and AI within societies, with a particular emphasis on gender and cross-cultural studies. She is the founder of the research lab NuMe, and her work on these topics began in the early 2000s. Leopoldina's favorite social robot is JOY ROBOT—a DIY (Do It Yourself) robot that emerged within the maker movement in Brazil in 2016, aimed at developing technology capable of engaging communities and assisting children in hospitals.

Natalie Friedman (Ph.D., Cornell Tech) is a UX Researcher at SAP (New York, USA). Her research focuses on human-robot interaction from a social sciences and design perspective. With a background in cognitive science, she observes in-field interactions between robots and people in many contexts including agriculture, hospitality, home, and public settings. Natalie has received an honorable mention award for her paper on designing clothing for robots at Designing Interactive Systems in 2021. In 2024, Natalie completed an art fellowship at the Interactive Telecommunications Program (ITP) at the New York University (NYU) working on a fashion show for robots where she is inviting artists to challenge gender roles through workwear for robots. She loves her electric toothbrush because it cleans her teeth!

Mafalda Gamboa (M.Sc./M.A., Chalmers University of Technology and University of Gothenburg) is a Ph.D. candidate in Interaction Design at Chalmers University of Technology in Gothenburg (Sweden). Her work focuses on design epistemology and first-person methods, building on her architecture, illustration, and photography practice. She is interested in feminist ways of knowing and re-telling stories surrounding technology through a mixture of ethnography and fabulation. Supported by her research work, she is also an accomplished and awarded pedagogue. She is currently funded to study flying robots and is reluctantly interested in understanding the ways drones are reconfiguring sociotechnical assemblages.

Dr. Zhenyu Gan (Ph.D., University of Michigan) is Assistant Professor in the College of Engineering and Computer Science at Syracuse University (New York, USA). His research is primarily dedicated to advancing the understanding of bipedal and quadrupedal systems, with a particular focus on gait analysis, gait transitions, and iterative learning control. Dr. Gan leads the Dynamic Locomotion and Robotics Lab, serves as the Form & Function Focus Group Leader at the Bioinspired Institute, and holds the position of Senior Research Associate at SU's Autonomous Systems Policy Institute. Notably, his favorite robot is Cassie, a bipedal robot by Agility Robotics, capable of walking and running akin to the Cassowary bird.

Joel Gn (Ph.D., National University of Singapore) is lecturer at the School of Spatial and Product Design, LASALLE College of the Arts (Singapore). His research lies in the intersections of media, design and popular culture, with an emphasis on the aesthetics and phenomenology of contemporary artifacts. An avid fan of anime and manga, Joel's favorite social robot is the hilarious and lovable Arale Norimaki from Akira Toriyama's *Dr. Slump*.

Tomasz Grzyb (Ph.D., SWPS University) is Dean of the Faculty of Psychology in Wrocław at the SWPS University (Poland) and President of the Polish Social Psychological Society. His main areas of interest are social influence and obedience toward authority. His favorite robot is WALL-E for a very simple reason: Although it looks like a pile of garbage, it still has feelings and thanks to them, it saves the world.

David J. Gunkel (Ph.D., DePaul University) is Presidential Research, Scholarship and Artistry Professor in the Department of Communication at Northern Illinois University (USA) and Professor of Applied Ethics at Łazarski University (Poland). He is an award-winning educator, researcher, and author, specializing in the philosophy of technology with a focus on the moral and legal challenges of artificial intelligence and robots. He is the author of more than 90 scholarly articles and has published 14 books, including *Thinking Otherwise: Philosophy, Communication, Technology* (Purdue University Press, 2007), *The Machine Question: Critical Perspectives on AI, Robots, and Ethics* (MIT Press, 2012), *Of Remixology: Ethics and Aesthetics After Remix* (MIT Press, 2016), *Robot Rights* (MIT Press, 2018), and *Person, Thing, Robot: A Moral and Legal Ontology for the 21st Century and Beyond* (MIT Press, 2023).

Katriina Heljakka (Doctor of Arts, Aalto University; Ph.D., University of Turku) is a researcher at the University of Turku (Pori, Finland). Her research focuses on toys and other playable media and machines as part of life-long play and playful learning. Her work on speculative toy fiction emphasizes the possibilities of artificial friendship and play machines that move us physically, cognitively, and emotionally. Katriina's favorite social robot is the "B*bot Ron" from the movie Ron's Gone Wrong. She owns a toy version of Ron but eagerly awaits future robots with 360-degree projection screens to arrive on the market.

Aike C. Horstmann (Ph.D., University of Duisburg-Essen) is postdoctoral researcher at the department of Social Psychology: Media and Communication, University of Duisburg-Essen (Germany). In her research she focuses on the social-psychological effects of human-machine-interaction, with an emphasis on the perception of and reaction to interactive technologies such as robots and virtual agents. Aike is still looking for her perfect robot match—however, there is definitely a strong bond with NAO after years of working together.

Dayeoun (Day) Jang (M.A., Hanyang University) is a Ph.D. student in the Media & Information department at Michigan State University (USA). She is interested in augmented body perception through connection with technological devices and/or experiences, especially avatars, virtual/mixed reality, artificial intelligence, and social robots. Before she started her academic journey, she was a co-founder of a start-up company making shape-changeable gaming controllers. Dayeoun likes social robots without specific purposes or that behave differently than they are programmed, such as Bastion from the *Overwatch* series and LOVOT, a pet robot made by Groove X.

Xiaoyu Jia (Ph.D., National Taiwan University of Science and Technology) is a lecturer in the College of Design and Innovation at Zhejiang Normal University (China). Her cross-disciplinary research interests include industrial and interaction design, human-robot interaction, and user experience research and design. Dr. Jia has won the 2012 Red Dot Concept Design Award and the 2014 IF Concept Design Award. In 2022, she presided over the ideological and political teaching project in Zhejiang Province. Dr. Jia's favorite robot is Baymax from the movie *Big Hero 6*.

Kevin Koban (Ph.D., Chemnitz University of Technology) is a postdoctoral researcher at the Department of Communication of the University of Vienna (Austria). He studies and publishes on interactive media, including mobile social media, social robots, and video games, with a focus on both beneficial and harmful ways of using them. He previously worked in the U.S. Air Force-funded project "Moral Agency in Robot-Human Interactions: Perceptions, Trust, & Influence" (MARIA) at Texas Tech University and is currently engaged in the European Research Council (ERC) project "Digital Hate: Perpetrators, Audiences, and

(Dis)Empowered Targets." Currently, Kevin's favorite robot is the Artificial Friend Klara from Kazuo Ishiguro's *Klara and the Sun* due to "her" magically illuminating understanding of the social world surrounding her.

Yoon Lee (M.A., Ewha Womans University) is a Ph.D. candidate at the S.I. Newhouse School of Public Communications, Syracuse University (New York, USA). Her research dissects user engagement with media across social media and virtual reality platforms, aiming to navigate and mitigate the complexities introduced by swift technological advancements. Her work aims to cultivate a symbiotic relationship between technology and its users. Yoon is the lab manager of the Newhouse Interaction Lab and a member of the Extended Reality Lab. Her favorite robot is Taekwon V, which embodies courage, teamwork, and the pursuit of justice—but most importantly, it has a great theme song!

Qingyu Liang (Ph.D., Beijing Jiaotong University). He is a visiting scholar at the School of Information Studies at Syracuse University (New York, USA). His research focuses on the future of work with the format of human-AI collaboration. His past and current work focuses on several key areas: the impact of AI on people's identities, AI ideation techniques design, and the factors that influence people's willingness to work with AI. Qingyu's favorite robot is Baymax in the movie *Big Hero 6*.

Roc Myers, (Colonel, USAF (Retired); M.S., University of Southern California) is founder/principal at Pertis (New Mexico, USA), a consultancy specializing in research, design and development of dynamic machine learning in complex, dynamic environments. His research focuses on developing synthetic cognition (SynCog) algorithms for collaborative, autonomous systems. He has 50 years' experience in research, development, and operation of advanced technology systems. His favorite robot is Juan San Ceros, a member of a team of SynCog robots in a dynamic and dangerous virtual world called Icos Island where they learn to collaborate, solve problems, avoid threats and outwit predators, and find resources to survive and thrive.

Kristine Nowak (Ph.D. Michigan State University) is Professor and Director of the Human Computer Interaction Lab in the Department of Communication at the University of Connecticut. Her research examines how humans use technology in ways that influence trust, learning, attribution, and perception. Her work has been published in several journals including the Journal of Computer-Mediated Communication, Media Psychology, and Technology, Mind & Behavior. She was awarded Fulbright's Fondazione con il Sud award in 2023, which allowed her to be a Visiting Professor at the University of Palermo and made the collaboration on this chapter possible. Her favorite robot is WALL-E because he spent his life trying to save the planet and didn't stop cleaning up after others even when he fell in love.

Uchenna Ogenyi (Ph.D., University of Portsmouth) is Lecturer at the School of Computing, University of Portsmouth (United Kingdom). His research focuses on machine learning and its applications in human-robot interaction and collaboration. He is part of the team that implemented the AiBle, an upper-limb rehabilitation exoskeleton robot project aimed at supporting stroke patients by providing advanced functionality to enable remote but active rehabilitation. The project was funded by the European Union, European Regional Development Fund. Among social robots, the Aibo robot holds a special place for him because of its lifelike behavior, emotional responsiveness, and capacity to learn make it into a genuine companion for its owners.

Sarah Rajtmajer (Ph.D., University of Zagreb) is Assistant Professor in the College of Information Sciences and Technology at The Pennsylvania State University (USA). Her research brings together machine learning, AI, and hybrid human-AI systems to understand how information encodes values like accuracy, objectivity, and privacy, and the trade-offs involved in managing healthy information ecosystems. Her work has been funded by the National Science Foundation, Defense Advanced Research Projects Agency, Office of Naval Research, and the U.S. Air Force Office of Scientific research. While Deep Blue inspired for possibility, Sarah's favorite AI incarnation was a 7-foot rolling "wingman" robot named Abe that she helped to build for the military during her post-doc. Abe was never quite Deep Blue, but the collaboration included a fully sentient neuroscientist, Frank, who later became Sarah's colleague and life partner.

Rafael Sousa Silva (B.A. Computer Science, B.A. Mathematics; Franklin & Marshall College) is a Ph.D. student at the Colorado School of Mines (USA). His research focuses on the implementation and evaluation of cognitive models of working memory applied to cognitive architectures for language-capable robots. He is a member of the 2023 HRI Pioneers cohort and served as the 2024 HRI Pioneers Sponsorship Chair. Rafa's favorite robot is one of the greatest social robots of all time, the one and only Kismet.

Jan-Philipp Stein (Dr. rer. nat., Chemnitz University of Technology) is Associate Professor at Chemnitz University of Technology (Germany), where he leads the department of Media Psychology. His research interests include human–machine communication, parasocial phenomena, and media effects on identity and the self. In his work, he often assumes a transdisciplinary perspective, for instance when examining users' relations to AI-powered robots from both psychological and philosophical perspectives. JP's favorite robot is a sassy one: Claptrap from the Borderlands videogame series.

Daniel Ullrich (Dr., LMU Munich) is Post-doc Researcher in the research group for Human-Computer Interaction at LMU Munich (Germany). His research revolves around social robots and smart spaces, with a special focus on trust, acceptance, (robot) personalities, and general interaction experience. His research paradigm is grounded on the premise that robots do not actually need to possess human-like skills—it is sufficient that humans assume they do. Daniel's favorite robot is TARS from the 2014 science fiction movie *Interstellar*, known for his recognizable personality, sarcastic remarks, and (barely) adjustable humor settings.

J. Nan Wilkenfeld (MBA/MA, UNC Charlotte) is a Ph.D. Candidate at the University of California, Santa Barbara (USA). She studies the increasingly interdependent relationships between humans and emerging intelligent technologies, focusing on interpersonal power dynamics between humans and their machine partners. She has worked on several grants from agencies such as the National Science Foundation and the U.S. Department of Defense. Nan's favorite social robot is Bender from Futurama as he is the most humanlike, not-idealized robot: Complex, flawed, rebellious, witty, introspective, and he enjoys a good adult beverage.

Tom Williams (Ph.D., Tufts University) is Associate Professor of Computer Science at Colorado School of Mines (USA). His research focuses on the effective, ethical, and equitable design of language-capable robots. He is the recipient of early career awards from the National Science Foundation, National Aeronautics and Space Administration, and the Air Force Office of Scientific Research, and was a Program Chair for the 2024 ACM/IEEE International Conference on Human-Robot Interaction. Tom's favorite social robot is Fresh Cut Grass, from Campaign 3 of *Critical Role*.

Elena Yifei Zhao (M.S., University of Southern California) is a Ph.D. student at Syracuse University (New York, USA). Her research focuses on game studies and media psychology, particularly how digital technologies have influenced components of our personalities and shaped our life outcomes.

Index

A

accountability to robotic machines, 258
action, or behavior, 238–239
actuators in social roboti c
 systems, 128–132
adaptability through changing
 clothing, 65–66
aesthetics
 color, role of, 64–65
 of machine bodies, 179–186
 social consequences, 63–70
affective loop, 88
affordances and play experiences, 108
agency, 197–204
Agent, assembled, 4
Aibo, 125
" Alexa's "brain," 202
Ameca, 39, 40
animalistic dehumanization, 260
anthropomorphized social robots, 66–67,
 79, 181, 188

ANYmal robot, 46, 49
artificial emotional intelligence, 89
artificial emotions, robotic models, 88
artificial general intelligence, 241–243
artificial intelligence (AI), 6
 AI-powered robots, 256
 AI-related techniques, 217–224
 misapplied notions of social cognition
  in, 232–233
 and robotics, 227
 social cognition, misapplied notions
  of, 232–233
 teammate, 218–223
artificiality of social robots, 185
artificial minds, 257
ASIMO, 25, 44, 46
(dis)assembling the Android, 1–6
ATRIAS, 44
attraction and repulsion, 182–183
audio sensors, 127
authority and influence, context of
 robots, 207–214
averted gazes, 37

B

battery power, 201
B*Bots, 111
BigDog, 46
biologically implantable
    microchips, 137–138
biometric sensors, 131–132
body language expression within robotic
    systems, 87
Buddhism, 138

C

TheCampaign to Stop Killer
    Robots, 265–266
Capek, Karel, 267, 273
Cartesian dualism, 260
ChatGPT, 260
clothing
    anthropomorphic forms and
        materials, 66–67
    changing, adaptability, 65–66
    and color, 63–70
    and color in context of robots, 63–70
    designing, and robots in tandem, 68
    human notions, 66–67
    robot updates, 67
co-construction of shared context, 230–231
cognition, 228–229
    architectures, 239
    psychology, 115–121
cognitive architectures, 239
cognitive psychology, 115–121
communicating (artificial) emotion in
    human-robot interactions, 88–90
contextualization, 229–230
convergence of form and being, 11–20
Cozmo, 14, 248
critical and cultural perspectives on touch
    and robots, 78
C-tactile afferents, 74

cuteness
    ethological factor, 180
    sociality, 180–182
cybernetics and organism
    (cyborg), 135–142
    acceptabilit y politics, 139–140
    broken-world thinking, 141
    lifetime relationships with entangled
        technologies, 140
    long-lived relations between technology
        and humanity, 141
cybernetic technology (CT), 135–137
cyborgian beings, 136

D

Dash toy robot, 106
Da Vinci Surgical Robot, 44
decision, 238
DeepMind, 241
deep reinforcement learning (RL)
    and generative AI, 239–240
defensive obligations of robots, 268
deictic gazes, 38–40
de la Bellacasa, Puig, 56
dependence, 180–182
descriptive language, 12
differentia specifica, 260
digitality, 167–172
digital worlds
    interactivity-as-demand
        model, 172–173
    social presence and electronic
        propinquity, 169
direct gaze, 35–37
distributed, integrated, affect,
    reflection, and cognition (DIARC)
    architecture, 120
distributed social robot, 148
distribution, 146
    social robot, 148
Dragonbot, 68

E

edutainment, 105
electric shock generator, 211
electromechanical actuators, 129
electronic propinquity, 168–170
emotional body language, 89–90
emotional displays, 257
engineering and design decisions, 95–98
episodic memory, 118
ERICA, 69

F

face-to-face (F2F) social interaction, 231
    human-robot shared context, 233–234
fictional robots, 79–80
Fisher-Price Smart, 106
Flobi, 24
flying robots, 53–60
frames, 157–162
    social and psychological
        effects, 160–162
    social robots' behaviors, 159–160
Frankenstein, 137
functional agency of robots, 199

G

gaze aversion, 37
gaze relations, 37
gendered robots
    gendering cues and human
        perception, 189–191
    for human understanding of
        gender, 187–193
genderless categorizations, 187
generative AI—systems, 260
gestalt-analytical approach, 12
gyroscopes, 132

H

Haraway, Donna, 137
Harmony model, 189
heimlich, 184
Huggable, 73
human-AI teaming, 218–221
human and computational models of
    memory, 116
human and machine agency, 197–198
human behavior and communication
    styles, 23
human-centered AI (HCAI), 218
human cognitive system processes, 12,
    13, 115
human diversity, 188
human gaze interactions, 40
human genders and robot gender
    cues, 191–193
human-human touch, 77
human-like emotion, 90–91
human–machine interaction
    (HMI), 256–257
human nature (HN), 183
humanoid robots, 27–28
    creation, 23
human-robot interaction (HRI), 23–25, 63
    communicating (artificial)
        emotion, 88–90
    social interface, 26
human-robot social interaction, 227
hydraulic actuators, 130

I

ideological frames, 137
images
    of robots in mass media, 158–159
    social and psychological
        effects, 160–162
information, 115–121

buffer, 120
inner-speech process, 98
innovations in processing capabilities, 97
integration, 238
intelligent minds in machines, 258
intelligent technologies, 202
Interactive AI, 241
interactivity-as-demand model, 172–173
Internet-connected toys (IoToys), 106
IoToy robots, 109
   with affective relations, 110–112

J

Jibo, 24

K

Kargu-2, 213
KASPAR, 86
Keepon, 86
KeJia, 145
Kinova Gen3, 69
Kirobo Mini, 184
Kismet, 23, 24
Kuratas, 14

L

labeled objects, 97
latent power of machines, 203
legal person, 276–277
legged animals and robots, gaits of, 47–48
legged mot ion controls, 45–46
legged robotics, 43–50
Leka robot, 86
lesser-evil justifications, 269
limb-environment interactions, 43–50
long-term memory, 116–119

M

machine bodies, aesthetics of, 179–186
machine vision and listening research, 80
mantling, 267
material agency, 198
material power, 201–202
Mavic 2 drone, 57
MAYA (most advanced, yet acceptable) principle, 152
mechanical gestures, 179
   and posture, 86–91
mechanistic dehumanization, 260
mediated-communication environments, 168
memory-centered approaches, 115
memory experiments, 210
memory systems, 115–121
mental capacities, 256
Milgram's studies and replications, 211
military robots, 267
mindless tools, 255
Mini-Cheetah, 46–48
misapplied notions of social cognition in artificial intelligence, 232–233
mobile robots, 201
moral agents, 258
moral and legal ontology for 21st century, 277–280
moral patiency, 259
Mori's model, 184
morphing, 147
Moxi, 210

N

natural language generation (NLG) systems, 234
natural person, 275–276
natural speech recognition systems, 98
noise in social robot communication, 99–100

non-player characters (NPCs), 167–174
normative approaches, 78

O

obedience research, 210
observation, 238
offensive objectives, 271
ontology, robots, 273–280
operationalizing touch for robots, 80
oral speech communication, 99

P

Pepper
  case study, 101
  default interactive gaze-seeking, 37, 101
Pepper Parlor concept cafe in Tokyu Plaza Shibuya, 34
perceptions of aliveness and deadness, 247
Personalities for Machinery in Personal Pervasive Smart Spaces (PerforM), 148
person/thing dichotomy, 278
physical and emotional experiences, 185
physical and perceptual systems, 74
physical movement and emotional bonding, 111
player avatar interaction scale (PAX), 170
player-avatar relations model (PAR), 170–172
playful affordances of future friends, 105–112
Pleurobot, 47
pneumatic actuators, 129
police robot, 213
political debates, 136
posture, 86
pragmatics of dressing, 69–70
preliminary ontological framework, 12
private speech, 98
propelling care, 58–59
proprioception module, 99
proximity sensors, 127–128

R

R2-D2, 85
research on human sex differences in human-robot interaction, 192
retroactive interference, 120
robot communication design, 95–103
robot gazes
  aesthetics and politics, 34–35
  communicating emotions, 34
  on human perception and behavior, 39
  non-verbal interpersonal communication, 35
  recognition, 35
  social and spatial information, 34
  technical and aesthetic elements, 35
  visual interrogation and demands, 34
robot gestures and postures
  mechanics of, 86–91
robot heads and faces
  characterizing, 24–25
  on HRI, 25–26
  mindful intentionality, 25
RoboThespian, 2
robotic agents, 111
robotic expressions of pain, 257
robot-initiated touch, 76–77
robot legs and feet
  shapes, 44–45
robot morphology
  adaptability and modularity, 15
  aesthetic design and thematic elements, 16
  manipulators and end effectors, 14–15
  materials and textures, 15
  mobility and locomotion, 14
  shaping perceptions, 13–16
  silhouette and body structure, 13–14
robot ontology
  form-function attribution bias (FFAB), 16–17
  new ontological category (NOC), 17–18
  physical forms and attributes, 16
robot power, 197–204

and agency, 200–204
robots
    biomimetic designs, 18
    clothing and color, 63–70
    constitutive technologies into, 237
    decision-making and situational awareness, 99
    design, 18
    emotional behavior, 88
    familiar humanoids, 18
    familiar humanoids, biomimetic designs and pre-robotic machines, 18
    heads or head-like features (*see* robot heads and faces)
    as in-between entities, 248–249
    life-death template, 249–252
    minimally invasive surgeries, 19
    patient outcomes, 19
    pre-robotic machines, 18
    as sociotechnical assemblages, 2–4
    teachable machines, 105
    thoughts and inner speech, 98–99
Roomba, 66
"room intelligence" (RI), 148
    characteristics, 148–149
    distributed social robot, 152
    operational elements, 151
    physical distribution and degree of control distribution, 152, 153
    prototypical sketching and experimentation, 150
    technical operationalization, 149

S

scholarship throughout human history, 247
self-defense, 259
sensing and movement abilities, 132
sensor-actuator synergy in robotics, 130–131
sensors
    in social robotic systems, 125–128
    types of, 75

short-term memory, 119–120
simplicity, 180–182
simulated emotional states, 85–91
sociable machines, 105–106
social agency of intelligent robots, 199–200
social and material agency of machines, 204
social and psychological functions, 137
social cognition, 228–229
    in artificial intelligence, misapplied notions of, 232–233
social communication, 98
social drones, 53–55
    design, 55
    dimensions of care, 55–59
sociality, cuteness, 180–182
social power of machines, 202–203
social presence, 168–170
social robot communication
    noise in, 99–100
    opportunities and design challenges, 102–103
social robotics, 23
    anthropomorphism, 179
    artificial agents, 179
    design and engineering, 95–98
    interactivity-as-demand model, 173
    locomotion for, 48–50
    player-avatar relations model (PAR), 171–172
    representation and design, 162
    social presence and electronic propinquity, 169–170
    tactility, 75
    technical systems of touch for, 74
    texture and tactility, 74
    with thermoception, 76
    touch systems, embrace orientation, 78–81
Sony AIBO robotic dog, 179
Sophia, 90
speech recognition systems, 98, 100
Spot, 85–91
StarlETH, 45

Star Trek universe, 146
study-session snapshots, 118
synergies of perception and
 movement, 125–132

T

tactile sensors, 73, 128
tactile systems, and texture of
 robots, 74–77
task-related expertise, 219
team dimensions, 222
teaming, 218–223
technologically-augmented toys, 106–107
texture, tactile perceptions of
 robots, 75, 77, 79
Theory of Core Knowledge, 117
threats to human distinctiveness, 259
3-D printed propeller guards, 58
TikTok creators, 136–137
toys, 106–107
traditional gendering through robot attire, 67–68
transformable robotics, 49

U

uniquely human (UH), 183
unpiloted aerial vehicles (UAVs), 213

V

valence-arousal-dominance (VAD), 88
vertical take-off and landing (VTOL)
 drone, 58
videogame avatars, 4
video games
 interactivity-as-demand
  model, 172–173
 player-avatar relations model
  (PAR), 171
 social presence and electronic
  propinquity, 169
vision-based sensors, 126–127

Y

YouTube creators, 136–137

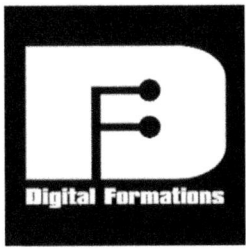

General Editor: **Steve Jones**

Digital Formations is the best source for critical, well-written books about digital technologies and modern life. Books in the series break new ground by emphasizing multiple methodological and theoretical approaches to deeply probe the formation and reformation of lived experience as it is refracted through digital interaction. Each volume in Digital Formations pushes forward our understanding of the intersections, and corresponding implications, between digital technologies and everyday life. The series examines broad issues in realms such as digital culture, electronic commerce, law, politics and governance, gender, the Internet, race, art, health and medicine, and education. The series emphasizes critical studies in the context of emergent and existing digital technologies.

Recent titles include:

Datafied Childhoods: Data Practices and Imaginaries in Children's Lives
Giovanna Mascheroni and Andra Siibak
Metaphors of Internet: Ways of Being in the Age of Ubiquity
Edited by Annette N. Markham and Katrin Tiidenberg
Media Distortions: Understanding the Power Behind Spam, Noise, and Other Deviant Media
Elinor Carmi

Making Our World: The Hacker and Maker Movements in Context
Edited by Jeremy Hunsinger and Andrew Schrock
Produsing Theory in a Digital World 3.0: The Intersection of Audiences and Production in Contemporary Theory – Volume 3
Edited by Rebecca Ann Lind
Ethics for a Digital Age, Vol. II
Edited by Bastiaan Vanacker and Don Heider

To order other books in this series please contact our Customer Service Department:

peterlang@presswarehouse.com (within the U.S.)
order@peterlang.com (outside the U.S.)

To find out more about the series or browse a full list of titles, please visit our website:

WWW.PETERLANG.COM

www.ingramcontent.com/pod-product-compliance
Ingram Content Group UK Ltd.
Pitfield, Milton Keynes, MK11 3LW, UK
UKHW021328180426
11947UKWH00017B/1499